Cosmopolitics I

CARY WOLFE, SERIES EDITOR

10 *Cosmopolitics II*
Isabelle Stengers

9 *Cosmopolitics I*
Isabelle Stengers

8 *What Is Posthumanism?*
Cary Wolfe

7 *Political Affect: Connecting the Social and the Somatic*
John Protevi

6 *Animal Capital: Rendering Life in Biopolitical Times*
Nicole Shukin

5 *Dorsality: Thinking Back through Technology and Politics*
David Wills

4 *Bíos: Biopolitics and Philosophy*
Roberto Esposito

3 *When Species Meet*
Donna J. Haraway

2 *The Poetics of DNA*
Judith Roof

1 *The Parasite*
Michel Serres

ISABELLE STENGERS

Cosmopolitics I

I. The Science Wars

II. The Invention of Mechanics

III. Thermodynamics

posthumanities 9

TRANSLATED BY ROBERT BONONNO

UNIVERSITY OF MINNESOTA PRESS
MINNEAPOLIS
LONDON

The University of Minnesota Press gratefully acknowledges the generous assistance provided for the publication of this book by the Hamilton P. Traub Press Fund.

Published by the University of Minnesota Press
111 Third Avenue South, Suite 290
Minneapolis, MN 55401-2520
http://www.upress.umn.edu

Library of Congress Cataloging-in-Publication Data

Stengers, Isabelle.
[Cosmopolitiques. English]
Cosmopolitics / Isabelle Stengers ; translated by Robert Bononno.
v. cm. — (Posthumanities ; 9-)
Includes bibliographical references and index.
Contents: I. The science wars, II. the invention of mechanics, III. thermodynamics
ISBN 978-0-8166-5686-8 (v. 1 : hc : alk. paper) — ISBN 978-0-8166-5687-5 (v. 1 : pb : alk. paper)
1. Science—History. 2. Science—Philosophy. 3. Science—Social aspects. I. Title.
Q125.S742613 2010
501—dc22

2010010387

Printed in the United States of America on acid-free paper

The University of Minnesota is an equal-opportunity educator and employer.

24 23 22 21 20 19 18 17 10 9 8 7 6 5 4 3

CONTENTS

PREFACE

How can we examine the discordant landscape of knowledge derived from modern science? Is there any consistency to be found among contradictory or mutually exclusive visions, ambitions, and methods? Is the hope of a "new alliance" that was expressed more than twenty years ago destined to remain a hollow dream?

I would like to respond to these questions by arguing for an "ecology of practices." I have constructed my argument in seven steps or parts, covering two separate volumes (this is the first).[1] Each of these seven books is self-contained and can be read on its own, but I hope that readers view individual books as an invitation to read the others, for the collection forms a unified whole. Step by step, I have attempted to bring into existence seven problematic landscapes, seven attempts at creating the possibility of consistency where there is currently only confrontation. Whether the topic is the nature of physics and physical law, the debate over self-organization and emergence, or the challenges posed by ethnopsychiatry to the division between modern and archaic knowledge, in each case I tried to address the practices from which such knowledge evolves, based on the constraints imposed by the uncertainties they introduce and their corresponding obligations. No unifying body of knowledge will ever demonstrate that the neutrino of physics can coexist with the multiple worlds mobilized by ethnopsychiatry. Nonetheless, such coexistence has a meaning, and it has nothing to do with tolerance or disenchanted skepticism. Such beings can be collectively affirmed in a "cosmopolitical" space where the hopes and doubts and fears and dreams they engender collide and cause them to exist. That is why, through the exploration of knowledge, what I would like to convey to the reader is also a form of ethical experimentation.

BOOK I

The Science Wars

1

Scientific Passions

How do the sciences force us to conceive of the world? What do they teach us about the possibilities of understanding it? According to Stephen Hawking, speaking with all the apparent authority of cosmological theory and as a descendant of Galileo, Newton, and Einstein, we will soon know the mind of god. John Wheeler, using quantum mechanics, claims that the universe itself, like everything that exists in space-time, owes its actual existence to the observer. Believers in the (strong) anthropic theory claim that science is leading us toward a different, but equally unsettling, conclusion: the end point of the universe is the production of those who describe it. This gives rise to the question of the durability of our cosmic vocation: what will become of mankind in a few billion years when the sun's resources are exhausted and the universe itself winds down? For the moment, however, we still don't know if quantum mechanics will allow Schrödinger's cat, enclosed in its infernal box, to die before the physicist condescends to open it, or if the entire universe will spin off multiple realities each time a measuring device produces one result rather than another. There is still ongoing debate concerning the possibility of the "stardust" we consist of achieving conscious experience: is consciousness an irreducible property, like space or time? Can it be fully explained in terms of the multiple cross-processing of information residing in the brain? Or, rather, is it based on quantum effects that have been amplified and stabilized in the brain's microtubules? Whatever the case may be, if thought can be reduced to the properties of circuits and neuronal systems, shouldn't we begin to treat our ideas about knowledge, the ego, consciousness, perception, and so on as fated to join the crystal spheres of astronomers, the phlogiston of chemists, or the animal spirits of physicians in the cemetery of prescientific theories?

It is said that the first step in the history of science was the break with myth, but equally important was the break with sophism. Rational discourse would, therefore, from its inception, designate its "others" polemically: the fictions that evade verification and defy argument, on the one hand, and the arguments that exploit the freedom—for those who have escaped myth—to prove a thesis (or its opposite), on the other. What of the historical sophists, apart from their role as outcasts, as the other of the philosopher, the friend of truth? How do myths function within the cultures in which they are an integral part? There is no need to raise such questions here, for terms like "myth" or "sophist," insofar as the sciences are concerned, serve as code words, always addressed to others, reminding them of the always renewable rupture. From this perspective it could be said that the sciences follow a narrow path, ever on the defensive against the powers of the imagination, which are satisfied with explanations and significations forged without constraint, and against the powers of rhetoric, which are satisfied with the ambiguity of language and the pretenses of proof.

In following this narrow path, are scientists really capable of balancing and theorizing the "larger questions" concerning the universe, its origin or finality, human thought, or humanity's role? If not, is it by again promoting the abstinence and proud humility that science must maintain in the face of the delicious temptations of ideology that we will be able to promote a harmonious and pacific collaboration among the hardworking "seekers of proof" extolled by epistemology?

In fact, the past and present of so-called scientific practices, as inventive as they may be, force those who study them to acknowledge that those qualities are always susceptible of turning into their opposite—narrow-mindedness and arrogance—as soon as those who are responsible for cultivating them are forced to position themselves against one another. If the landscape of practice currently provides the impression of coherence, it is one of generalized polemic. Cold or hot, depending on circumstances, it is expressed as contemptuous disinterest, attempts at annexation (for example, that long-awaited moment when a "rational pharmacology" will finally enable us to design "scientific" drugs), even dramatic proclamations, where a contested practice links its fate to that of humanity as a whole (the criticisms of psychoanalysts who warn of the threat presented by the rise of pharmacological psychiatry). This polemic is embodied statically in our universities, where every discipline has its own territory, its experts, its criteria, and where the reassuring fiction of collegiality prevails, one whose only point of agreement is the disqualification of the "nonscientific." A polemic embodied much more dynamically by the "large-scale operations"[1] of mobilization, conquest, and hierarchization that structure the landscape of the scientific disciplines.

Thirty years ago, the person who wrote those lines, then a novice philosopher, still believed in the exemplary role physics could play once it affirmed the possibility of transforming the scope and significance of its function as a model for other forms of knowledge—a function it has served ever since the origin of the modern sciences. *Order out of Chaos: Man's New Dialogue with Nature,* which I coauthored with Ilya Prigogine in 1979, showed how some of the most fascinating statements made by physics, particularly the reduction of the distinction between past and future—"time's arrow"—to a mere question of probability, far from conferring upon physics a quasi-prophetic function, disclosed its fragility, the impassioned adventurousness of its character. At that time Ilya Prigogine and I wrote: "In any event, as far as physicists are concerned, they have lost any *theoretical* argument for claiming any privilege, whether of extraterritoriality or of precedence. As scientists, they belong to a culture to which they in their turn contribute."[2] In *Entre le temps et l'éternité,* we again stated that "the search for coherence among forms of knowledge has been the connecting thread in this essay. . . . We cannot discover such coherence as if it were a truth that transcends our history, whether that history leads us to truth or has lost its original connection to it. We can only construct it within this history, from the constraints that situate us but which also enable us to create new possibles."[3] However, it is much easier to announce the good news that the prophetic utterances of physics have changed and now reflect a world that is temporally asymmetric rather than symmetric, chaotic or bifurcating rather than deterministic, capable of self-organization and not inert and static, than to face the bemused smile of readers confronted with the idea that physicists are capable of telling them what kind of world they live in. That is the lesson I needed to learn. In a sense, for the third time, I intend to rework this notion of coherence and to do so by confronting the question of the relationship between the "passion for truth" characteristic of the scientist, and which marked both *Order out of Chaos* and *Entre le temps et l'éternité,* and the question of a possible peace, a humor of truth.

One possible objection is that the lesson was obvious and should have been evident to any philosopher worthy of the name. The very title of the book, *Order out of Chaos,* didn't exactly serve as an example of a denial of prophetic emphasis. And whenever it attempted to turn physics into a "poetic attentiveness" to nature, didn't it already—even though we had specified that poetics was to be understood in the etymological sense of "maker"—encourage scientists to range outside the narrow and austere pathways that defined them, with respect to myth as well as the precariousness of discursive proof? And am I not once more in the process of making the same mistake? Why speak of the humor of truth when the association between "science" and "truth" is now suspect? Shouldn't I acknowledge that it is the responsibility of critical thought, which teaches each of us the limitations of our respective approaches, to promote methodological peace?

I have to acknowledge that the ideal of peace through a rejection of the ambitions and passions that the critic condemns is not my goal. What's more, it seems to me that this ideal is one whose history leads us to doubt its relevance. After all, if there is a turning point in what is referred to as modern science, wasn't it Galileo's rejection of the eminently rational compromise offered by Cardinal Bellarmin? If the astronomers had been in agreement, the heliocentric doctrine would have been recognized as "true," but it would only be so relative to the questions and calculations of the profession. Indeed, one could also claim that the great narrative of the Copernican revolution, which celebrates the destruction of the ancient cosmos, with the Earth at its center, and its substitution by an acentric universe in which the Earth is merely a planet, was by no means necessary. For the Earth-as-planet is less a substitute for the Earth-as-center than a supplement; it is a reference point for new questions, new practices, and new values, but does not produce genuinely scientific answers to age-old questions. But Galileo's rejection of the Jesuit proposal must be heard. The Earth-as-planet is not a simple professional hypothesis, it asserts a truth that no methodological ban will be able to limit. Can we ask that Galileo's heirs endorse the ascetic rejection he himself refused to make?

One might reply that this backward movement is illegitimate, as the period in question was one of conflict, a time when more than just the relative positions of the Earth and the Sun were at stake. Galileo was defending freedom of thought in the face of clerical dogma, that is, the possibility of genuine critical thought. Methodological criticism can only take place in a pacified world, a world where the right to conduct research and the absence of revealed knowledge are recognized. Galileo's heirs no longer need, or should no longer need, weapons of questionable merit to conquer a territory that is recognized as their own.

Let's look at another example. In 1908, a time when religious dogma was no longer a threat to physics, the physicist Max Planck initiated the excommunication of his colleague Ernst Mach, whom he proclaimed guilty, through his historical-pragmatic conception of physics, of weakening the faith in the intelligible unity of the world. For Mach, physical references that appeared to refer to a world that existed independently—absolute space and time, atoms, and so on—had to be eliminated and replaced by formulations that tied physical laws to the human practices with which they were indissolubly connected. In contrast to this critical approach, Planck would affirm the necessity of the "physicist's faith" in the possibility of achieving a unified concept of the physical world. Without that faith the source of inspiration that had enabled minds such as Copernicus, Kepler, Newton, and Faraday to carry out their work would dry up.[4]

Planck was the first to explicitly position physics within the context of *faith* rather than austere rationality, a faith that had now become an essential component of the physicist's *vocation,* and to correlatively affirm that the practice of physics was not just another kind of science. Planck did not actually deny the general plausibility of Mach's description, he rejected it *for physics.* Physicists *must be able* to speak of the "world" or "nature" independently of the operational and instrumental relationships that, for Mach, were the only source of theory's legitimacy. Without that, how could physicists have dared claim that energy is conserved and that it was already conserved before life on Earth even existed, that is, before a human was able to conceive of it? How could they have felt authorized to claim that the laws of gravity would continue to govern the movements of celestial bodies after the destruction of the Earth and all its inhabitants? In order to be able to produce such statements—the culmination of modern physics—Planck states that the physicist must be able to believe that even an "inhabitant of Mars," or any other intelligence in the universe, can produce their equivalent. The differentiation established by Planck, based on which he defined the "physicist's vocation," does not juxtapose opinion and rational practice but affirms the privilege of physics. In doing so, he connected the inspirational needs of physicists with a twofold hierarchy: one for the "realities" with which we deal, with physical reality being the only "real" one, and that of our rational knowledge, with physics at the summit.

Here, Planck created what Gilles Deleuze and Félix Guattari refer to as a "psychosocial type."[5] Planck's physicist is not a portrait, one we might want to compare against the original. His role is to serve as a "marker," to function as a reference whenever physicists discuss their work, its meaning, and the scope of their theories. And the faith that inhabits Planck's physicist cannot be assimilated to a type of ideological overload indifferent to what one might recognize as strictly scientific challenges. While the theme of the physicist's vocation may reflect a strategy of hierarchization, it cannot be reduced to such a strategy in the sense that it could be understood in purely human, social, political, or cultural terms. Planck is not inventing a means of differentiating physics from the other sciences; he states, he literally "*cries out*" against Mach the *fact* of that difference. He celebrates the conservation of energy but he is himself the product of the event engendered by the statement of that conservation, the victim of the power it seems to confer on the physicist: the power to talk about the world independently of the relationships of knowledge that humans create.

As such, the impassioned vocation of the physicist affirmed by Planck is part of the present, of the identity of physics transmitted to physicists, with which they identify in turn. And that vocation serves as a reference not only in "external" discourse on the rights and claims of physics but within strictly technical controversies that underlie concepts considered fundamental by physicists. It is in itself a vector and ingredient of history. The "physicist" whose commitment it heralds is, for better or worse, an integral part of the very construction of the theoretical claims of twentieth-century physics.

It seems to me that the impassioned commitment of physicists is bound to resist criticism precisely because it has been forged in opposition to critical thought, like that of Mach,[6] and because an active component of the history physicists inherit and which they learn to extend can be found in its reference to the scandalous creativity of a physics that rejects the limits proposed by critical rationalism.

Yet, we may very well wonder whether this vocation, and with it the scientific faith that serves as an obstacle to methodological peace, are not part of a past only traces of which remain in the present, with those being mostly media related. Clearly, a certain type of "prophetic" physics exists today. But, if we must speak of physics, wouldn't it be preferable to approach it from the viewpoint of the new undertaking known as "big science"? International financing, the construction of large-scale instruments, management of an experiment over a period of several years, the organization of large numbers of colleagues, the division of labor: these are the kinds of practical questions that preoccupy "cutting-edge" physicists today far more than the "ideal" question of the physicist's vocation. Can't we take advantage of this situation, which clearly illustrates that, regardless of its vocation, physics is confronted with the same kind of difficulty faced by every mega-enterprise threatened by bureaucratization and autism, and forget about this outdated mess of arrogant pretensions?

It is an objection we need to take very seriously. A plausible future is within sight in which there will obviously be scientists, but they, as more or less competent employees, will no longer be distinguished from anyone else who sells their labor power. That this perfectly plausible future already serves to disqualify interest in the impassioned singularity of scientific practices may appear to be an appropriate response to the arrogance of their claims. In *The Invention of Modern Science* I wrote of the connivance of the so-called modern sciences with the dynamics of redefinition that singularize this delocalized, rhizomatic power known as capitalism.[7] We can see the genial hand of capitalism in this complicity, the source of its most formidable singularity: its parasitic nature. While capitalism has destroyed many practices, it also has the ability not to destroy those it feeds on but to redefine them. So-called modern practices are affected by this parasitism, which gives them an identity that weakens any ability to resist their subjugation, pits them against one another, and leads them to condone the destruction of practices whose time has come. Wouldn't it be fair if scientific practices, which have to a certain extent benefited from the dynamic of redefinition that destroyed so many others, were to experience the same fate?

However, this vindictive morality, no matter how appealing it may be, is not one I share. Its promulgators will always have good reasons for their verdict, but this verdict will be delivered repeatedly, without risk, and situates them in a monotonous landscape littered with similar reasons for disqualification. Where then can we situate, in our present, a "cause" capable of resisting the accusation of compromise and able to teach us to resist, along with it; a cause that we can acknowledge to be free of complicity, having resisted not through some historical contingency predicated on "not yet," but through its own resources, the dynamics of capitalist redefinition? If learning to think is learning to resist a future that presents itself as obvious, plausible, and normal, we cannot do so either by evoking an abstract future, from which everything subject to our disapproval has been swept aside, or by referring to a distant cause that we could and should imagine to be free of any compromise. To resist a likely future in the present is to gamble that the present still provides substance for resistance, that it is populated by practices that remain vital even if none of them has escaped the generalized parasitism that implicates them all.

Consequently, it is the "living" physicist I need to consider, not the one who will snicker at the romantic dream pursued by her science and which a harsh reality will have destroyed. I do not want to take advantage of the process that would replace the generalized polemic among practices with the creation of an instrumental network where each discipline would have no other identity but that of a data generator that marks its position in the network in question. I want to resist this process. This presupposes betting on the possibility of different dreams for physicists and other modern practitioners. Therefore, it is the anxiety that continues to occupy the physicist at CERN that I want to confirm and celebrate, and not the likelihood of the cynical laugh that ushers in the abandonment of the dream and the redefinition of the physicist as a cog in some more or less extravagant large-scale undertaking.

"The *diagnosis* of becomings in every passing present is what Nietzsche assigned to the philosopher as physician, 'physician of civilization,' or inventor of new immanent modes of existence," wrote Deleuze and Guattari.[8] The challenge they lay out could equally be my own: to diagnose the "new immanent modes of existence" our modern practices may be capable of. This also implies the possibility of "psychosocial" types activated by a desire for truth that would not require them to claim—as it did in the case of Planck and Ernst Mach—access to a truth that transcends all others.

The Invention of Modern Science culminated in the apparently paradoxical figure of "nonrelativist sophists," of practitioners capable of claiming that "man is the measure of all things" and of understanding the statement "not all measurements are equivalent" as an imperative, to make sure we have made ourselves worthy of addressing what we claim to measure. Those sophists who are not satisfied with the mere acknowledgment of the relativity of truth but would affirm the truth of the relative[9]—what I refer to as the humor of truth—would then be equally capable of reworking the meaning of the relationship that identifies science and struggles against opinion and myth. For—and this is the central thesis of *The Invention of Modern Science*—while the "struggle against opinion" is vital to the so-called modern sciences, that struggle has nothing to do with matters of principle: the opinion against which a science is invented is not opinion in general. It is opinion created with reference to the invention itself, to the possibility of a new "measurement," of the creation of a new way, always local and relative, of differentiating science from fiction. That is why I have tried to highlight the difference between the event constituted by the creation of a measurement and the directive embodied in the reduction of this event to an illustration of the right and general obligation to subject all things to measurement. This difference can be stated in political terms, and it would then correspond to the difference between the politics constitutive of the sciences and a general politics of power. Yes, scientific practices, and in particular theoretical–experimental practices, are vulnerable to power but, no, this vulnerability cannot be confused with fatality. This difference can also be stated in terms of a "mode of existence": the sciences do not owe their existence to the disqualification, with which they are identified, of so-called "prescientific," or nonrational, knowledge.

Yet the possibility of other identities for the sciences, as I tried to bring out in *The Invention of Modern Science,* is not sufficient for the operation of "diagnosis." A true diagnosis, in the Nietzschean sense, must have the power of a performative. It cannot be commentary, exteriority, but must risk assuming an inventive position that brings into existence, and makes perceptible, the passions and actions associated with the becomings it evokes. What I want to make perceptible are the passions and actions associated with a peace that is not one of method, that does not demand that those it involves reject the specific passion for truth that allows them to think and create.

Naturally, the act of diagnosis must not be confused with a mere political project. It is not a question of constructing a strategy that hopes to inscribe itself as such in our history and which, in order to do so, must take into account the interests and effective relations of force without which no claim, no objective, no alternative proposal would have meaning. If it were a question of strategy, the undertaking would be part of a genre that has demonstrated its ability to survive its own absurdity: it would position me in line with those—and they are legion—who are convinced that everyone's future is governed by conditions that they themselves are responsible for establishing.

The diagnosis of becoming is not the starting point for a strategy but rather a *speculative* operation, a thought experiment. A thought experiment can never claim to be able to constitute a program that would simply need to be put into application. With respect to scientific practices—as elsewhere—such experiments have never had any role other than that of creating possibles, that is of making visible the directives, evidences, and rejections that those possibles must question before they themselves can become perceptible. And unlike the thought experiments that are part of scientific practices, these possibles are not determined, and what is at stake is not the creation of an experimental mechanism for actualizing and testing them. The diagnosis of becomings does not assume the identification of possibles but their intrinsic link with a struggle against probabilities,[10] a struggle wherein the actors must define themselves in terms of probabilities. In other words, it is a question of creating words that are meaningful only when they bring about their own reinvention, words whose greatest ambition would be to become elements of histories that, without them, might have been slightly different.

2

The Neutrino's Paradoxical Mode of Existence

I would like to return to the point where I approached the question of the "physicist's vocation." It is indeed in terms of mystification that Mach criticized the reference to atoms, and to absolute space and time. Seen from the perspective of the references accepted at the time concerning the opposition between an authentically scientific practice and one not subject to the exigencies of scientific rationality, Mach was "right" and Planck was well aware of this. He knew he was associating the "physicist's vocation" with what, following Marx, should be referred to as mystification: the transformation into "the properties of things themselves" of something that, according to Mach, should be subject to experimental practice and, Marx would have added, to its corresponding social relations. It is this that may have triggered the violence of Planck's reply, the accusation that Mach was a "false prophet": we recognize false prophets, he said, by the fruits of their prophecies, in this case the predictable death of physics.

But it is Émile Meyerson, the philosopher of science, who best understood the violence of the rejection by physicists of the "rational" translation of their quest that had been proposed by critical philosophy. For this emphasized a generalized *presentation* that contrasted the passion for comprehension with the ascetic reading offered by epistemology. In the beginning of Meyerson's first great book, *Identity and Reality* (1907), he notes the difference between a "law" and a "cause." Although ordinary epistemology took pride in following Hume in its critique of causality, which should, rationally, be reduced to empirical regularity (where the law would define the rule), Meyerson showed that scientists are not satisfied with such regularity, even if it allows them to predict and control. On the other hand, every time a causal hypothesis has led us to assume a nature capable of explaining itself, it has, he claims, exercised its hold over physicists. The nature of this hypothesis—that atoms collide according to Cartesian laws, that they are attracted to one another in a Newtonian sense, that they are replaced by energy as understood by Ostwald, or by disturbances in the ether, or by a pure physical–mathematical formulation—is of little importance. What is important, for Meyerson, is the construction of an "ontological" reality that could explain what we observe and could do so, moreover, by reducing change to permanence, by demonstrating the *identity* of cause and effect. Reason anticipates and expects identity, that is, the discovery of some permanence beyond an observable change, and it does so even when the possible realization of its ambition for identification would have paradoxical consequences. "Let us suppose for a moment that science can really make the causal postulate prevail; antecedent and consequent, cause and effect, are confused and become indiscernible, simultaneous. And time itself, whose course no longer implies change, is indiscernible, unimaginable, non-existent. It is the confusion of past, present, and future—a universe eternally immutable. The progress of the world is stopped. . . . It is the universe immutable in space and time, the sphere of Parmenides, imperishable and without change."[1]

From Meyerson's point of view, the idea of a stable separation between science and metaphysics is a vain pursuit: "Metaphysics penetrates all science, for the very simple reason that it is contained in its point of departure. We cannot even isolate it in a precise region. *Primum vivere, deinde philosophari* seems to be a precept dictated by wisdom. It is really a chimerical rule almost as inapplicable as if we were advised to rid ourselves of the force of gravitation. *Vivere est philosophari.*"[2] Every time the possibility of understanding arises, no matter how bold and speculative, it benefits from a favorable a priori: scientists have a *propensity* for considering that possibility to be true; it seems "plausible" to them. For Meyerson, plausibility is neither aprioristic nor empirical. Unlike a Kantian aprioristic judgment, it may be refuted by experiment, but it nevertheless exerts a seductive power on the mind of the scientist, just as it does on "common sense" in general, that no empirical knowledge alone is capable of justifying.

Because it exists nature can bend to the requirements of the causal postulate only partially. It manifests itself, therefore, in its "irrationality," in the resistance the effort at identification always runs up against. This points to the great difference between the history of a science such as physics, where the general and invincible tendency of the human mind to identify is reflected in the risk and creativeness generated by resistance, and other undertakings that are satisfied with plausibility. To state that the physical brain must obviously explain thought, for example, is to embrace a "plausible" statement in the Meyersonian sense, and the difference between the static flatness of this statement and the beauty of Einstein's vision is explained by the poverty of the constraints the first will have to satisfy, as well as the consequences that will have to be verified. Neither aprioristic nor empirical, such a statement can indifferently associate itself with any aspect of neurophysiological research.

I have dwelt on Meyerson's thesis at some length because it quite accurately describes the challenge that lies before me. Had I accepted his claims, my problem would be solved. It would be pointless to investigate the meaning assumed by the "physicist's faith" at the turn of the last century or the failure of the various criticisms leveled at physics. The prestige of the theories that lend to physics the allure of metaphysics, the hierarchy of the sciences, as well as the hierarchy that characterizes physics and divides it into "fundamental physics" and "phenomenological physics," limited to the study of observed behaviors, would be self-explanatory. And it would be robustly self-explanatory because no critical disclosure of any kind could modify what would then possess the allure of fatality. There would be nothing left to do but offer some slightly incantatory praise for the "risk" that characterizes the difference between the physicist's "faith" and the vacuity of common sense whenever it gets mistaken for science.

But being faced with a challenge does not mean that I have the means to refute a description like the one provided by Meyerson. On the contrary, I regard such a description as terrifyingly *plausible,* much more plausible than those that view physics as a project for domination or control. Learning to resist this very Meyersonian plausibility, learning not to "identify" physics, whatever the temptation, with a metaphysical common sense that would explain its successes and its excesses, is an attempt to implement a different idea of philosophy, one that I have already referred to specifically as "speculative" in the sense of a struggle against probabilities.

The possibility of a "non-Meyersonian" solution affects the past much less than the future. Especially the future of the relationship between what we call science and what we call philosophy. If Meyerson had been right, those relationships would be stable, the scientist repeatedly producing statements that would appear to be a presentation of the "real in itself," the philosopher adopting a critical position, reminding us, now as before, of the illegitimate character of those statements, the illusions on which they are based. And they will continue to provide "fictions of matter," which fictionalize physical reality as able to explain life or consciousness, and to exhibit the irrepressible "fetishization" of the beings constructed by the experimental sciences. Hasn't the molecule born in the laboratories of physicists, to the great displeasure of those rational chemists who denounced it the way we denounce fetishes, now been offered, in the form of DNA, to the public at large as the key to human salvation, the holder of the—purely genetic—secret of human destiny?

"We must destroy our fetishes!" This is the slogan that provides critical thought with an all-purpose foundation. "Common sense is fetishistic, irrepressibly fetishistic, and the destruction you demand is none other than your own, of the passion that is the life of your intellect." This was Meyerson's reply, which Planck would probably agree with. And Planck might also add, as Einstein did ("the real incomprehensible miracle is that the world turns out to be comprehensible"), that, where physics is concerned, fetishistic faith is de facto confirmed. But what those two antagonistic positions have in common is that they both seem to know a great deal, a bit too much in fact, about fetishes, about the way they function, about the "common sense" that all mankind is said to share, about the irrepressible tendencies all cultures are said to manifest. In this sense, Meyerson, Planck, and Mach are indeed *modern,* as the term is understood by Bruno Latour, in that, regardless of their conflicts, they belong to a culture whose curious singularity is that it defines relationships to what are globally referred to as "fetishes" in terms of belief, although they are prepared to disagree over whether such belief is indispensable or not.

When Mach attacks the fetishes that feed on thought, he demands that the decisive break that defines modernity be recognized, and maintained, in the face of the temptations of "regression." "Men" must not only recognize that their practices are an integral part of the referents they cause to exist but that those referents refer only to those practices. They renounce any syntax that might be directed toward an autonomous reality. Naturally, such renunciation confirms the ability and vocation of modern practices to disqualify all other practices, which do not define themselves as antifetish, but it is this renunciation that Planck refuses to accept on behalf of physics, and Meyerson, on behalf of common sense itself. However, from the point of view of my hypothesis of the possibility of a "a nonrelative sophist," such a refusal is inadequate. It is not fetishistic belief that needs to be defended, but rather a "cult of fetishes" in all their diversity, modern and nonmodern.

This is the decisive step taken by Bruno Latour in his *Petite Réflexion sur le culte moderne des dieux faitiches,*[3] and it is Latour I will follow here so I am able to approach Planck's rejection in terms other than those of an unjustifiable faith, justified in fact. What Planck defends against Mach is not only the physicist's "faith" in a vision of the metaphysical-physical world, it is also—and I am gambling that it is primarily—the fact that the beings fabricated by physics may nonetheless be referred to as "real," endowed, no matter that they are "fabricated," with an autonomous existence: "*factishes,*" as Latour calls them.

To abandon antifetishistic critical thinking docs not imply acceptance of Planck's position as such, or acknowledgment that physics uncontrollably tends toward metaphysics. It is to introduce the *possible* ambiguity of its position. The theme of "faith," which Planck makes a condition of physics, could be understood as a protest by someone who feels forced by an antifetishistic adversary to reject what is, for him, the greatness of his undertaking. The theme of belief—"leave us our fetishes; obviously, we're the ones who create them, but we need to believe, we vitally need to believe in their autonomy"—would reflect the strength of the modern antifetish position: Planck would have no way, other than in terms of belief or faith, to describe what in his eyes makes physics valuable, that which it cannot abandon. But based on this hypothesis, what Planck wants to affirm is, primarily, that the creatures physics brings into existence possess, as their *constitutive attribute,* the power to legitimately claim an autonomous existence. Without the impassioned yet demanding tests that have verified that legitimacy, they would not exist. As for the large theme of necessary belief in a vision of a unified world, far from reflecting an irrepressible Meyersonian tendency, it would imply that modern antifetishism, which destroyed the words Planck needed, replaced them with a claim that has all the seductiveness of a war cry. The reference, not to the autonomy of "physical beings" (atoms, electrons, neutrinos, etc.), but to an autonomous *world* that would ensure the unique authority of physics, allowed Planck to shift from defense to offense, to counter the authority of critical thinking with the authority of the tradition of physics as a whole.

My (speculative) interpretation means that the question of the vocation of the physicist can be addressed in terms that are no longer "general purpose" but inherent in the art of fabricating "factishes," which singularizes physics. Planck was able to defend this singularity only by joining it to "belief." But isn't the need to affirm such belief associated with the definition of modern practices as "antifetishistic?" And with respect to physical beings, doesn't the possibility that their claim to autonomy can be understood noncritically then suggest a new approach to the theme of the physicist's vocation? In other words, isn't it possible that the "factishes" passionately constructed by physicists, were they recognized as such, might maintain, with the references constructed by other forms of knowledge, relationships that are not hierarchical and polemical?

I have in mind here the creation of a "psychosocial" physicist whose practice would require her to consider, and whose practice would make possible, at the same time and coherently, these two apparently contradictory propositions: that the neutrino is as old as the period in which its existence was first demonstrated, that is, produced, in our laboratories, and that it dates back to the origins of the universe. It was both constructed and defined as an ingredient in all weak nuclear interactions and, as such, is an integral part of our cosmological models.[4] Consequently, it can serve as the subject of propositions that make it a product of our understanding and others that make it a participant in a cosmic history that is said to have led to the appearance of beings capable of constructing such understanding.

I choose the neutrino because it exemplifies in a particularly dramatic way the paradoxical mode of existence of all those beings that have been constructed by physics and that exist in a way that affirms their independence with respect to the time frame of human knowledge. The demonstration of the existence of an entity such as the neutrino obviously has nothing in common, as Meyerson showed, with the production of a general law based on observable and reproducible regularities.[5] It has nothing to do with the modesty of a simple description resulting from the activity of methodical and critical observation, that is, an activity that would boast of finally ridding itself of parasitic passions that paralyze rational inquiry. The neutrino sweeps aside this apparent modesty. It denies the idea that the products of science present no problem other than that of knowing why humans have, for so long, allowed themselves to be swept up by their passions and deceived by their illusions. And it does so in two complementary ways. On the one hand, it is a quintessential example of an object that is difficult to observe, for its primary attribute is to be susceptible only to interactions that occur very rarely: the devices that enabled it to attain its status as existent imply and assume an enormous number of instruments, interpretations, and references to other particles that have already come into existence for human knowledge, and, inseparably, a tangle of human, social, technical, mathematical, institutional, and cultural histories. Moreover, it is even more "charged" because the existence of this genuinely phantom particle, which ignores walls and barriers, had been postulated, for theoretical-aesthetic reasons of symmetry and conservation, long before the means for "detecting" it were created. However, once the means were created and once it demonstrated its existence under the required conditions, the neutrino existed with all the characteristics of a real "actor," endowed with properties that also enable it to act and explain, autonomous in relation to the detection device that caused it to bear witness to its existence and which is now nothing more than an "instrument." For this was the vocation of the existence it was endowed with: the proofs upon which the legitimacy of that existence within physics depended were supposed to give the physicist the power to claim that the neutrino had existed for all time and in all places, and that the effects that make it observable and identifiable by humans are events that demand to be understood as ingredients not of human history but of the history of the universe.

The neutrino is not, therefore, the "normal" intersection between a rational activity and a phenomenal world. The neutrino and its peers, starting with Newton's scandalous force of attraction, bind together the mutual involvement of two realities undergoing correlated expansions: that of the dense network of our practices and their histories, that of the components and modes of interaction that populate what is referred to as the "physical world." In short, the neutrino exists simultaneously and inseparably "in itself" and "for us," becoming even more "in itself," a participant in countless events in which we seek the principles of matter, as it comes into existence "for us," an ingredient of increasingly numerous practices, devices, and possibles. This apparently paradoxical mode of existence—in which, far from being at odds, as is the case in traditional philosophy, the "in itself" and the "for us" are correlatively produced—is indeed the one targeted by experimental practice in the strong sense, the one whose triumph is measured by its ability to bring into existence *factishes* that are both dated and transhistoric.

To follow Latour in calling "factishes" those beings we fabricate and that fabricate us, from which the scientist (or the technician, via different modes) "receives autonomy by giving [them] an autonomy he does not have,"[6] does not confer upon them any identity other than the fully developed identity they get in physics. That is why it is important to speak of factishes and not fetishes, for I am not trying to establish a general theory of fetishes, which would never be more than the pseudopositive counterpart of their general condemnation. On the contrary, beginning with the question of what allows the practitioner to claim that the beings she fabricates exist autonomously, it entails posing the problem of the distinct modes of existence of the beings we bring into existence and that bring us into existence. As will be shown, the distinctions begin within physics itself and their number increases whenever we try to understand the impassioned interest in new artifacts capable of being referred to as "living" or even "thinking."

There is nothing consensual or pacific about the "factishes" we bring into being. Recognizing them as irreducible to a critical epistemology or to the kind of "objects" philosophy likes to contrast with "subjects" is not at all synonymous with pacification and coherence. But to recognize them as such may function as a *proposition* addressed to their "creators." Such a proposition, while affirming the singularity of their practice as being creative, with no obligation to the great narrative that contrasts myth and reason, is not limited to ratifying what they insist on seeing recognized. It is an *active* proposition that can involve them in sorting out whatever it is they claim, and especially to consider superfluous the claim to the power of disqualification. In other words, factishes propose a humor of truth. They create the possibility of a divergence between two themes that are frequently coupled: transcendence and assurance. Yes, the creature transcends its creator, but this is no miracle but an event whose production polarizes the work of the creator.[7] No, the produced transcendence does not guarantee membership in a transcendent world, or the availability of that world as such as a reference for judgments or operations of disqualification or annexation. Factishes are a way of affirming the truthfulness of the relative, that is, a way of relating the power of truth to a *practical event* and not to a world to which practices would merely provide access.

The factishistic proposition does not claim a neutrality that would be accepted by all. It invites the physicist and other constructors of factishes to differentiate the conquered-fabricated-discovered autonomy of their creatures from the unengendered autonomy of a world waiting to be discovered. But it also reflects a trust, which the neutrino does not necessarily justify and which is not even specifically addressed to it but which concerns all those existents produced in experimental laboratories. To gamble on the possibility of the humor of a truth that acknowledges its fabrication is to commit oneself to a future where irony does not triumph: these existents will not dissolve inside a mournful and sempiternal network of compromise and negotiation that, once deciphered, would lead to the conclusion that they are fabrications pointing toward a routine of human, all too human, negotiation.

In one sense, I'm trying to reenact the scene between Planck and Mach. Mach's criticism does not allow the physicist to "present himself," to define his vocation, because the words offered require that he deny his passion for truth. Is the factishistic proposition able to do so? Can the vision of a "physical world" defended by Planck lose its seductiveness? Can it be recognized as a "default" response, accepting, for want of anything better, the adversary's references, the opposition between antifetishistic rationality and an irrational but fecund faith? Can factishes free physicists from a mode of presentation that encloses them in an alternative that is somewhat vulnerable to irony: either invoking a faith that would lead them forward the way a carrot leads a donkey, or laying claim to the successes of physics in order to affirm that it is really on its way to achieving its quest for the world's truth, for penetrating "the mind of god"? It is not up to me to decide.

In any event, the touchstone of my undertaking is much less the fabricators themselves than the way in which they are, or might be, present among us.

The sciences, as they are taught, that is, as they are presented once their results are unlinked from the practices of science "as it is practiced," do not have a meaning that is appreciably different from a religious engine of war, pointing out the path to salvation, condemning sin and idolatry. And it is not by appealing to an improved "scientific culture" that this problem is going to be resolved—the problem of the mode of existence among us of neutrinos, genes, fossils, and other scientific creatures. That such a culture is what is always missing, the thing whose absence is always invoked, whose existence would be a kind of panacea, without anyone being able to say what it might consist of (because the majority of scientists are, apparently, the first to lack this well-known culture), is a good reflection of the ghostly existence of what is being invoked. A ghost is not always lacking in power, however. In some cultures its appearance forces its members to think, connect, act.

In our culture the sempiternal return of the great theme of the necessary adjunct of "conscience," without which, apparently, science would be "the ruin of the soul," commits us to nothing, because what is asked is unclear. No practices exist, akin to those used by others to heed what insists and construct a message, a message that would create a difference. In our case, it would make connections and add new questions to those asked by scientists. In other words, we are haunted by the necessity of scientific culture although our practices do not provide it with the means to exist.

The manner in which the neutrino and other scientific

factishes "are presented" to those who do not share in their production can become a cultural question only if that culture is actively dissociated from "information," from the possession of "cultural knowledge." An awareness of the history of the neutrino's creation and the problems to which it responded cannot prevent its existence from being generalized into a "neutral" fact, that is, both an *authenticated* fact that everyone "should" be familiar with if they are to be modern citizens, and an *available* fact, which anyone may pick up and use for their own purposes. The question of knowing how the neutrino's existence is, could be, or will be celebrated does not find an answer either in the willingness that recalls, under no obligation, that the sciences are human works, or in the irony that recognizes the work behind the fact.

That the struggle not to forget the multiple components of the event that caused the neutrino to exist seems endless and hopeless does not reflect a "psychological" difficulty (humans prefer to believe than to understand) or an "epistemological" question (the context of the justification takes precedence over the context of discovery). It reflects the fact that the "discovery" of the neutrino is not an event likely to interest "mankind" as such. The neutrino does not mark a step along the path that leads "mankind" from ignorance to understanding; it owes its existence to the fact of having fulfilled what Latour calls a very demanding set of "specifications," of having satisfied very specific proofs, which allow "specific people," the community of its fabricators, to forget the avatars of its fabrication, to celebrate its existence "in itself." If something is to be celebrated or must force others to think, it is not the neutrino but the coproduction of a community and a reality of which, from now on, from the point of view of the community, the neutrino is an integral part. Such an event has yet to deserve interesting others. The cultural traditions that are not antifetishist cultivate such an interest. They know how the constructors of fetishes need to be addressed, what can be expected of them, why they should be feared. To consider the social, cultural, and political presence among us of the highly specific communities formed by the constructors of factishes may be a way of "materializing" the ghostly reference to a "scientific culture" that is always lacking.

The "acculturation" of the neutrino is, therefore, a practical question, inseparable from the relationships that need to be developed with those who brought it into existence, those whose proofs it satisfied. Other than that of a "neutral fact," the neutrino's identity will find stability only in a network of relationships through which new "immanent modes of existence" for our practices are invented. The touchstone of the factishistic proposition, and more specifically of what I am trying to do with it, is not to convince scientists but to bring about a transformation of the interests that identify them. And this, of course, is to be understood in the radically indeterminate sense authorized by the concept of interest: the way in which what one does interests others, that is, becomes an integral part of the present of others, or "counts" for others, does not conflict with the way in which one is interested in what one does oneself, but is an ingredient of it. Who is interested, how can one be interested, at what price, by what means and under what constraints—these are not secondary questions associated with the "diffusion" of knowledge. They are the ingredients of its identity, that is, the way in which it exists for others and the way in which it situates others.

3

Culturing the *Pharmakon*?

There are certain questions that, while they have resonance throughout the history of philosophy, assume particular significance in a given period. The question posed by the sophists is one of them and I want to address it explicitly, to prevent any possible misunderstanding.

The historical sophists were treated with opprobrium by philosophers, and were referred to as the philosopher's other: they were the ones who bartered the truth, who claimed to heal the city's woes without first obtaining knowledge of good and evil, who exploited the shadows and appearances of the "cave" rather than seeking the veridical light that reveals things in their proper guise. They were men who relied on opinion, changing and malleable. It's possible, of course, to return to that point in time to demonstrate the unfairness of this portrait or to "save" certain sophists from the judgment that has condemned them all. But it is not up to me to denounce their condemnation as an outright fabrication or adopt a position of indifferent objectivity that affirms historical neutrality. The question of our relationship to the sophists is not closed. Even more than the poet, who was also chased from the Platonic city but has since been reintegrated into an honorable civic category, the sophist, vector of lucidity or creator of illusion, doctor or soul thief, continues to trouble us.

The problem posed by the sophists is not dependent on any intrinsic quality that might be attributed to them but rather on their lack of it, that is, precisely the *instability* of the effects used to qualify them. One might even state that the sophist embodies this instability more than he produces it, and the recurrent comparison of the sophist with the *pharmakon,* a drug that may act as a poison or a remedy, clearly reflects this. The lack of a stable and well determined attribute is the problem posed by any *pharmakon,* by any drug whose effect can mutate into its opposite, depending on the dose, the circumstances, or the context, any drug whose action provides no guarantee, defines no fixed point of reference that would allow us to recognize and understand its effects with some assurance.[1]

The question of the *pharmakon* is not unique to the tradition that begins in Greece with the exclusion of the sophists. Every human culture recognizes the intrinsic instability of certain roles, certain practices, certain drugs. Tobie Nathan notes that if there is something unique about the West, it is its "confidence" in the doctor or psychotherapist, attributing to them the transparent desire to "heal" us (whatever the definition of heal might be). Other peoples are well aware that although he is able to heal, the therapist can also destroy:[2] the individual who manipulates influence can be savior or sorcerer. The instability of the *pharmakon* is not our specific problem. What does seem to make us unique, what the exclusion of the sophists, in its own way, seems to illustrate, is the intolerance of our tradition in the face of this type of ambiguity, the anxiety it arouses. We require a fixed point, a foundation, a guarantee. We require a stable distinction between the beneficial medicament and the harmful drug, between rational pedagogy and suggestive influence, between reason and opinion.

The contemporary scene is literally saturated with the "modern" heirs of Plato. Each of these heirs denounces his "other," just as the philosopher denounced the sophists, accused them of exploiting that which he himself had triumphed over. They include not only the heirs of Plato, but those philosophers who, following the sophists, were used as an argument to demonstrate the need for a foundation.[3] What, in Plato's text, can be read as a network of analogies isolating the terrible instability of the sophist–*pharmakon* has today split into a number of "modern practices" (scientific, medical, political, technological, psychoanalytic, pedagogical) that have been introduced, just as Platonic philosophy in its time, as disqualifying their other—charlatan, populist, ideologue, astrologer, magician, hypnotist, charismatic teacher.

It is possible—and tempting—to do to modern practices what was successfully done with Plato, who was shown by erudite readers to entertain an uneasy relationship to the sophists he denounced. Just as the sophists used Plato to promote their arguments, we can show that the question of relation, which traditional therapists were expert in, endures as an enigma in the very heart of medicine,[4] and that scientific demonstrations always imply an element of persuasion, although they claim to discriminate between objective proof and subjective persuasion. We can conclude, then, that the *pharmakon* refuses to be excluded, it inhabits the heart of the fortresses that are supposed to protect us from its instability. But, once again, we are limited to competing for the best, the most lucid form of "truth telling" without bringing into existence other ways of telling. And this "truth telling" locks us into a setting whose only horizon is what *we* call *pharmakon.* For the sophist, condemned for his exploitation of malleable and docile opinion, does not provide us with access to some generic definition, which is to say, robust, resistant to the contingency of circumstance. He himself is a contemporary of Plato, an inhabitant of the Greek city, where the question of politics was raised, the question of titles that authorized participation in governing the city's affairs. Similarly, the charlatan so-called modern medicine denounces is a "modern" charlatan, not the representative of what one would call "nonmodern" therapeutic practices.[5] There is nothing neutral about the definition of the insistent figure of the *pharmakon* as a symptom in the heart of whatever tries to distinguish itself from it; it is *our* definition, the one we have constructed by constructing the practices that have disqualified and, therefore, transformed, if not destroyed, the traditional ways in which this instability we associate with the *pharmakon* was managed.

This detour through the sophist and the *pharmakon* obviously amplifies what I have called the "factishistic proposition" and enables us to better identify the challenge. For if the question I want to present is that of the "presence among us" of physicists and other constructors of factishes, it seems obvious that this question is doubly pharmacological. First, because if we have to speak of "factishes" with respect to some of our productions, it is to the extent that these productions are intended to resist the pharmacological accusation. The neutrino, the atom, or DNA can claim that they "exist" autonomously with respect to the people who constructed them; they have overcome the proofs intended to demonstrate that they were not just fictions capable of betraying their author. They exemplified the experimental achievement: "the invention of the power to confer on things the power of conferring on the experimenter the power to speak in their name."[6] However, this threefold power is eminently limited for it is not warranted by an exterior fixed point, a generalizable definition of the difference between scientific statement and opinion or fiction. Once the neutrino, the atom, or DNA move away from the very specific site, the network of labs, where they achieved their existence, once they are taken up in statements that unbind existence, invention, and proof, they can change meaning and become the vectors of what might be called "scientific opinion"—scientific factishes have a very pharmacological instability.

It is with respect to this instability that we can formulate the question of the "nonrelativist sophist" capable of what Bruno

Latour would call the "cult" of factishes, where cult refers to a celebration of the event that brings a new being or a new method of measurement into existence, and to culture, the practice of establishing relationships. The pharmacological instability of our factishes, their terrible ability to feed the real obsession with the oppositional differentiation that makes us unique and repeatedly leads us to assign them a power they do not have, the power to disqualify practices and questions that don't concern them, already corresponds to a kind of cult. Polemics and disqualification are relationships and a component of practices. The "physicist's vocation" defined by Planck, although it appears to isolate him, contains the very opposite of that isolation, namely, the construction of a position of judgment that gives the "physical world" the power to transcend all other realities. The question of the "sophist" capable of celebrating and cultivating the event that constitutes the creation of a factish is new only because it responds to a new problem: *all cults are not equal.* That is why I want to risk qualifying the problem as "ecological."

The advantage of such an approach stems from the fact that the term "ecology" has a dual meaning, "scientific" and "political." Whatever the interdependence among populations of living beings may be, it can be called "ecological" in the scientific sense by its association with the concerns and research practices of scientific ecology. By analogy we can characterize the population of our practices, as such, as an ecological situation, regardless of the "immanent mode of existence" of each member or the nature of the contribution represented by the existence of other members for them. In contrast, for those ecologists whose commitment falls within a political register, not all "ecological" situations are equal, especially when they include members of the human species among their protagonists. Ecological practice (political in the broad sense) is then related to the production of values, to the proposal of new modes of evaluation, new meanings. But those values, modes of evaluation, and meanings do not transcend the situation in question, they do not constitute its intelligible truth. They are about the production of *new relations that are added* to a situation already produced by a multiplicity of relations. And those relations can also be read in terms of value, evaluation, and meaning.

There is, then, no substantive difference between the ecological situations studied by ecologists and those that strive to bring into existence the struggles conducted in the name of "ecological values," just as there is no substantive difference between the values, evaluations, and meanings created whenever a parasitic relation is transformed into a symbiotic relation, or when a parasite that destroys its host too efficiently is eliminated, and the values, evaluations, and meanings at the heart of ecological debates. In fact, there is hardly an ecological situation on Earth where the values attributed by humans to different "products" of nature haven't already contributed to the construction of relationships among nonhuman living beings. The only singularity of political ecology is to *explicitly* assert, as a problem, the inseparable relation between values and the construction of relationships within a world that can always already be deciphered in terms of values and relations. Which both changes nothing and changes everything, as is the case whenever what was implicit becomes explicit.[7]

Another advantage of the reference to "ecology" is that it refers to questions of process, namely, those likely to include disparate terms. Ecology can and should, for example, take into account the consequences, for a given milieu, of the appearance of a new technical practice just as it does for the consequences of climate change or the appearance of a new species. Following a logic of equivalence or intentionality, in each case the consequences do not reflect a "cause," nor can relations themselves be separated from the temporal regime of their entangled coexistence.[8] If there is one thing political ecology has had to learn from scientific ecology, it is that we should abandon the temptation to conceive of nature as submissive, manipulable, assimilable to some "raw material" on which we would be free to impose whatever organization we choose.

Ecology is not a science of functions. The populations whose modes of entangled coexistence it describes are not fully defined by the respective roles they play in that entanglement, in such a way that we could deduce the identity of each on the basis of its role. This role is by definition "metastable," that is to say not guaranteed against some possible instability. It is the product of "*bricolage,*" all we can say of which is that it "works more or less," and not of a calculation whose economy and logic would have to be disclosed. Correlatively, interdependent populations do not make a "system" in the sense that they could be defined as parts of a large whole. The point of view that allows us to describe the relative coherence between their respective modes of coexistence must itself interweave multiple timescales and issues. For example, a very rare species of bat appears to play no more than an insignificant role in the tropical forest of Puerto Rico. However, it has been found that its role is in fact crucial for the forest. After a hurricane, bats of this species are, unlike the others, incapable of flying away. Forced, at the risk of their lives, to survive where they are, the bats in this way contribute to the devastated forest's ability to recover.[9] By analogy, we can state that if the population of our practices presents, for me, the problem of a coherence that is not one of generalized polemic, a producer of arrogance and vector of submission, this coherence should have nothing in common with the coherence that authorizes a unitary point of view from which the role assigned to each participant can be deduced.

Ecology is, then, the science of multiplicities, disparate causalities, and unintentional creations of meaning. The field of ecological questions is one where the consequences of the meanings we create, the judgments we produce and to which we assign the status of "fact," concerning what is primary and what is secondary, must be addressed immediately, whether those consequences are intentional or unforeseen. Human societies are always susceptible to producing a justification for what they undergo, of transforming their inventions into norms, and forgetting the price paid for their choices. However, because of the construction of questions and knowledge produced by ecological practices, a new kind of memory has come into being, a memory of the unintentional processes that in the past were able to bring about the disappearance of cities, empires, or civilizations, and of the ravages caused by our simplistic industrial, and even "scientific," strategies (the "DDT strategy"). And this memory is now part of the present. In that sense, we can say that our present is cultivating the growth of "pharmacological knowledge," a science of processes where good intentions risk turning into disasters, and in terms of which no action has an identity independent of the whole that stabilizes it but causes it, under certain circumstances, to change its meaning.[10]

The "ecological" perspective invites us not to mistake a consensus situation, where the population of our practices finds itself subjected to criteria that transcend their diversity in the name of a shared intent, a superior good, for an ideal peace. Ecology doesn't provide any examples of such submission. It doesn't understand consensus but, at most, symbiosis, in which every protagonist is interested in the success of the other for its own reasons. The "symbiotic agreement" is an event, the production of new, immanent modes of existence, and not the recognition of a more powerful interest before which divergent particular interests would have to bow down. Nor is it the consequence of a harmonization that would transcend the egoism of those interests. It is part of what I would refer to as an immanent process of "reciprocal capture," a process that is not substantively different from other processes, such as parasitism or predation, that one could qualify as unilateral given that the identity of one of the terms of the relation does not appear to refer specifically to the existence of the other.[11] The specific "strategies" of mimetic defense employed by the caterpillar refer to the "cognitive" abilities of the bird that threatens it, but it seems that for the bird the caterpillar is just one kind of prey among others.[12] The definition of the parasite includes a "knowledge" of the means to invade its prey, but this prey appears to *simply* endure the parasite's attack. Both the caterpillar and the parasite exist in a way that affirms the existence of their respective other, but the opposite does not appear to be true—at least as far as we know at present. In contrast, we can speak of reciprocal capture whenever a *dual* process of identity construction is produced: regardless of the manner, and usually in ways that are completely different, identities that coinvent one another each integrate a reference to the other for their own benefit. In the case of symbiosis, this reference is found to be positive: each of the beings coinvented by the relationship of reciprocal capture has an interest, if it is to continue its existence, in seeing the other maintain its existence.

The concept of "reciprocal capture," like all those that bring to mind the stability of a relation without reference to an interest that would transcend its terms, allows us to emphasize the consequences of the ecological perspective I intend to adopt, primarily the lack of relevance in this perspective of the customary opposition between fact and value, the first referring to the order of "facts," the second to a purely human judgment. Whenever there is reciprocal capture, value is created. Naturally, scientific field ecology can rely on the stability of the situations it studies when producing representations and an evaluation of those situations. But once human practices come into play, the ecological perspective cannot rely on such stability but, on the contrary, communicates directly with the question of the pharmacological instability associated with *pharmaka* in general, and with the factishes that create and are created by our practices in particular. The question of the identity of a practice would then have to be answered not by a static diagnosis but by a question of "value" and "value creation," that is, the ecological question of what "counts" and "could count" for that practice. In other words, "politically" reclaiming factishes does not imply their submission "in the name of political values" but can be addressed through the immanent question of the way in which each practice defines its relationship to others, that is to say, "presents itself" to those others.

As a result, the perspective of an "ecology of practices" requires that we do not view "value" as that "in whose name" something can be imposed or must be accepted. Only humans on Earth act "in the name of values" and contrast them with "facts." But, and this holds true for humans as well as nonhumans, the *creation of value* cannot function in this register of opposition. The invention of a practice or the creation of a relationship of capture are part of what Félix Guattari, in *Chaosmosis,* has called "axiological creationism": a new constellation of "value universes" "that are detected as soon as they are produced and which are found to be already there, always, as soon as they are engendered."[13] Just as this creation cannot operate, among humans, in the name of values, it cannot, when it designates nonhuman beings, be associated with an explanation of a scientific nature, especially an interpretation that would simply make it the result of Darwinian selection. The concept of "value" as I use it here, and as Félix Guattari uses it in *Chaosmosis,* on the contrary, introduces the question of what we presuppose every time we give selection the exclusive power to explain. The (Guattarian) "creationist" perspective celebrates the existence of every given type of being that specifically poses the question of what counts for its mode of life. Axiological creationism concerns the production of existence for everything for which existence implies a "gamble," a risk, the creation of a point of view about what, from then on, will become a milieu.[14]

Can the creation of factishes be understood from the

vantage point of reciprocal capture? Yes, certainly, in the sense that here we have the coinvention of a being and those whose requirements it has satisfied: the neutrino exists for physicists and, somewhat differently, the physicists exist for the neutrino. Their definition now includes the questions and speculations authorized by the existence of the neutrino. The benefit of reciprocal capture is to enable us to understand this "for" in the strong sense of the coconstruction of identity rather than in the weak sense, which could be reduced to a simple convention.

To this distinction between strong sense and weak sense corresponds the usual meaning of the two terms "constructivist" and "relativist." Today, the respective use of these terms is far from fixed. I use the term "relativist" to qualify the reductionist position (the neutrino would *only* exist relative to the physicist) and the term "constructivist" to qualify the position that affirms the event, the twofold, correlated creation of a factish and its builder. One argument in favor of this choice, although somewhat arbitrary, is that historically the statement that our judgments are relative has always corresponded to a critical perspective, while the term "constructivist" now implies, in certain uses, an affirmative but risky perspective.[15]

According to the meaning I have given it, the constructivist ambition requires that we accept that none of our knowledge, none of our convictions, none of our truths can succeed in transcending the status of a "construction." It requires that we affirm their historical immanence and that we take an interest in the means invented, and in the authorities invoked, to establish their claim to a stability that would transcend history, taking those means and authorities as constructions that are added to the first. But the constructivist ambition does not require—quite the contrary—that we yield to the monotonous refrain "it *is only* a construction," as if an all-powerful truth were at stake, apparently the only one to escape the relativity it proclaims, thereby authorizing the widespread ironic parasitism of all the others. If constructivism, by its own criteria, has any truth other than that of the effects it produces, it does not in itself have the power to justify any derision, even tacit, in the face of the event that constitutes the stabilization of a practice, a knowledge, or a style. By analogy, ecology has accustomed us to consider that no species can be said to be "adapted" independently of a successful bet on the relationships it might have with other species and their shared environment. Biologists have had to stop producing general definitions that transcend situations. But this has not dampened their interest in the "success" referred to by the adjective *adapted,* their ability to celebrate the singular assemblage realized by every invented stability. On the contrary, their practice has become increasingly singular as they become more sensitive to the variety of modes of ecological stability.

The reference to reciprocal capture, moreover, has the advantage of helping us resist the temptation to confuse ideas and practices. A very current confusion. For example, the originality of Thomas Kuhn's concept of "paradigm" was to show us how the members of a scientific discipline learn to recognize and *treat* problems. This question of treatment has most often been forgotten, assimilated into the very conventional notion of a "vision of the world," with its equally conventional correlate of a silent world, one that allows itself to be indifferently deciphered and interpreted according to whatever ideas are prevalent at the time.[16] The contrast between practice and idea is crucial in that it stands in opposition to their hierarchical articulation: the idea conceived as vision would precede, inspire, and command practice, thus defined as application, a simple implementation. The idea so conceived is by definition limitless, capable of being freely extended, unaffected by the petty constraints of its particular "applications." It can encounter no resistance except that of other, antagonistic ideas. Its job is to rule, and it has no other challenge than the elimination of whatever serves as an obstacle to its rule. Correlatively, the theme of an "ecology of ideas" has the defect of mistaking beings who are self-sufficient, not put at risk by the world, for actors. That is why it culminates, quite conventionally, in the expectation of a conversion. As if by miracle, ideas would become "tolerant" of one another, capable of coexisting angelically, of lucidly recognizing that they are, in fact, "only ideas." As for the ecology of practices, it must state what it is that differentiates practice and idea: practices cannot, any more than living beings, address a silent world, the docile substrate of convictions and interpretations; their mode of existence is relational and constrained, not hallucinatory or visionary; their avatars do not refer to a more general authority for whom they would be a local translation, but to a here and now they fabricate and which makes them possible.

However, the notion of reciprocal capture is too broad to grasp the specificity of factishes. It encompasses everything that inscribes itself in a history and contributes to it, regardless of the scale of that history. It lends itself to a correlated historicization of what we distinguish as know-how or competence on the one hand, artifact or instrument on the other, starting with language itself and its most elementary act, the reciprocal capture of the sound uttered by "me" and of the word that "I" am not alone in having heard spoken, and including every intellectual technology.[17] In contrast, the (speculative) question I raise is not about the ecology of practices in general (how, for example, reading has changed oral practices or how the computer will alter the practice of writing) but about the problem presented to the ecology of practices by "modern" practices, marked by their Platonic heritage, creative of beings who owe their existence to the fact that they have satisfied proofs demonstrating that they were not merely fictions, dependent on their author. Even pedagogues recoil—fortunately—at the ambition of introducing, for the benefit of parents, a "rational" method of learning what is still referred to as our mother tongue. And large-scale technological constructions, like the subway, are designed to indifferently welcome anyone, whether stock traders or worshipers of Gaia, the only assumption being that they can read a word or a map. To grasp the specificity of factishes, the concept of reciprocal capture, which applies to subway riders as well as to the manufacturers of factishes, must be qualified so that this equivalence can be broken.

4

Constraints

How can we qualify a transversal concept like reciprocal capture without losing what it is that makes it most precious, namely, the fact that it shifts our attention and the focus of our questions? While conventional philosophical problems such as subject and object or doctrine and method always communicate with questions of right, legitimacy, and therefore the ability to disqualify as well, the concept of reciprocal capture emphasizes the event, an "It works!" that belongs to the register of creation. The criteria used to specify "what" works, the interests and problems reciprocal capture responds to, do not precede the event itself. Therefore, it is a question of qualifying the "It works!" while retaining this *abstract* character, but without turning it into a response to a problem that would condition the criteria to which a solution should respond before producing the solution.

I want to turn now to the notion of constraint in order to construct the specific landscape where reciprocal capture would allow the question of modern practices to be asked. Here, the notion of constraint has nothing to do with a limitation, ban, or imperative that would come from the outside, in other words, would be endured, and everything to do with the creation of values that I associate with the event of reciprocal capture.

Moreover, the meaning of "constraint" should be understood abstractly in a way that distinguishes it from "condition." Unlike conditions, which are always relative to a given existent that needs to be explained, established, or legitimized, constraint provides no explanation, no foundation, no legitimacy. A constraint must be satisfied, but the way it is satisfied remains, by definition, an open question. A constraint must be taken into account, but it does not tell us how it should be taken into account. It gets its meaning only in the process of coming into existence, thereby allowing the condition explaining that one thing rather than another has come into existence to be formulated a posteriori.

The notion of constraint invites us to situate the question of reciprocal capture in a landscape in which what should be satisfied is on the order of "holding together with others." The crucial importance of this question—once the question of innovation in the field of science and technology arose—has become one of the great lessons of the Parisian sociology of networks.[1] To follow the way a new idea "materializes" is first and foremost to follow the operations of recruitment and alliance that will produce the real "material" of innovation, the heterogeneous set of those who agree to be shaken up, modified, interested by it. Each of these could be said to "present its conditions," but materialization is a real history; for the talent of innovators is to transform conditions into constraints, in other words, not to submit to existing relations of force but to rework the implications, at least partially. It is after, and only after, the new set of relationships among all parties—human protagonists, technological devices, nonhumans, and so on—has been stabilized that we will be able to identify the factors explaining innovation (stakeholders, satisfied needs, reliability, profitability, etc.). In short, why and how and in what sense the innovation "works." The notion of constraint allows us to differentiate the subway rider and the producer of factishes, in the modern sense. It is possible to reduce the activity of "taking the subway" into a routine individual act precisely because the subway "works." But the construction of what we call "subway," and the construction of the identity of its user as well, corresponds to an eminently collective practice, which had to take into account eminently heterogeneous constraints that tied together eminently disparate protagonists.

However, the notion of constraint, in its turn, needs to be specified if it is going to allow me to ask the questions that interest me, for these imply a differentiation between "scientific" practices and "technological-industrial" practices. It goes without saying that science, technology, and industry are interrelated, even when we are dealing with a so-called pure science. Every scientific "reality," whether that of the neutrino or a fossil, is dense with technology, and technological devices in turn refer to the dense network of industries that produce them in series and mobilize them by assigning them new roles, new meanings, new challenges. But the ecological question I am constructing does not address this issue but, rather, that of knowing if it is appropriate, or if it is still appropriate, to distinguish between the two types of creation (scientific and technological). Of course, the way in which the distinction is usually produced, based on the opposition between a disinterested, autonomous knowledge and a finalized, circumstantial knowledge, pleads for the idea that there would be an advantage in abandoning it.[2] While this conclusion is plausible, what interests me is not the plausibility but the capacity to resist and invent.

To announce that the human fossil and the neutrino are part of the same mode of scientific existence is already highly contentious from an ecological point of view because the passion of reconstituting our heritage from scattered hominid fossil debris has little to do with the passion of subjecting the neutrino to the proofs on which its scientific existence depends, and a great deal to do with questions that are much older than

Darwinian science. But in this case the distinctions must be constructed and largely belong to the future.[3] In contrast, to state that the fossil, the neutrino, the "human genome" project, and the development of techniques of artificial procreation are part of one and the same enterprise is to assume the responsibility of ratifying some common characteristic, the thing that reflects their collusion with power. For only a reference to power can suggest what the critical demystification ratifies, namely, that the neutrino and the artificially fertilized female egg are both equally products of "technoscience." The neutrino and the fertilized egg do not share the same mode of existence at all, for their coming into existence did not face the same kinds of challenges.

Naturally, neither the techno-industrial "It works!" nor the scientific "It works!" is more or less disinterested, more or less autonomous than the other. All the same, they are not identi fiable once we try to understand them from the point of view of the constraints they must satisfy before being allowed to make history with us, and of the way this "us" with whom they make history is defined. If they appear as identifiable, if, for example, molecular biologists can announce that genetics has gotten a "second wind" by becoming genetic engineering,[4] we should not view this as the confirmation of a normal "technoscientific" becoming but as the announcement of a problem that demands attention and should raise multiple questions. What happened to "scientific" genes? Did they somehow disappoint the hopes of biologists? If that is the case, if the large-scale program initiated by biologists like Jacques Monod and François Jacob in the late 1960s—moving seamlessly from bacteria to the mouse—has met a roadblock, isn't it with respect to that roadblock that the redefinition of the practices of molecular biologists should be described? How did genes "mutate," how did they become the subject of an undertaking that claims to be related to the practices of engineers? How was the network organized to include those who, in order to heal, diagnose, prevent, modify, exploit, wished to refer to these new genes? The term "genetic engineering" signals a transformation that is both practical and professional, and the story of this transformation must accentuate the contrast in order to establish a difference—apparently insignificant yet crucial—between a disquieting history and an inflexible logic.

"How," asks Latour, "could we be chilled by the cold breath of the sciences, when the sciences are hot and fragile, human and controversial, full of thinking reeds and of subjects who are themselves inhabited by things?"[5] How can we be frightened of the "scientific gene," that factish so filled with human passion, so charged with fragile ambition, and so disputed because of the ambitions it has failed to satisfy? But how can we not fear the neutral gene, the common alibi of a thousand and one undertakings, unkillable because free to redefine itself according to the whim of situations that give it a thousand and one practical identities, freely available because it merely imposes on those who refer to it the need to create a link, no matter how tenuous, to the existing network? How can we not struggle against the confusion among the different ways of "creating" genes that, for sure, have accompanied them ever since their speculative origins, but have taken a new turn ever since the "second wind" of molecular genetics freed genes and biologists from many of the constraints associated with their claim "It works!"?

Comparing the challenges innovations must meet, depending on whether they involve technological-industrial or scientific productions, has nothing to do with a hierarchization that would allow scientists to demand for the sciences a freedom refused to the "applications" of their discoveries. It is a matter of distinguishing between the autonomous modes of existence of factishes that are fabricated in very different ways. The meaning of this distinction between modes of existence is the difference that distinction could make were it to become an element in the way in which corresponding existents are presented, that is, are present among us. The manner in which the sciences and "technosciences" are presented today cannot be judged either as veridical or as false or ideological, for it cannot be judged on the basis of an identity that would predate them. The manner in which they are presented is a part of their identity, like the production of all relations. For that reason it is the target, par excellence, of strategies of power, which need to smooth over differences in order to redefine constraints in the service of a logic both multiple and unanimous. I, on the contrary, intend to accentuate those differences.

Making those differences explicit means taking them into account when questioning how to address what is thereby differentiated. Until now I have used the term "practice" as if its meaning were self-evident: the practice of the physicist, the doctor, the technician, and so on. I now want to put it to work, for what needs to be made present, to be encouraged to present itself, is not the activity of an individual or the product of that activity. It is the ingredient without which neither that activity nor this product would exist as such. Individuals are not—here and to the extent that they interest me—isolatable, and the questions associated with the product of their activity cannot—again to the extent that it interests me—be separated from the constraints its coming into existence had to satisfy, the way questions associated with a meteorite or a thunderbolt, which comes from elsewhere and imposes itself where it will, can. The individuals who interest me are not qualified either in psychological terms or by gestures that can be identified in isolation: anyone who could, upon being shown images made in a chemistry laboratory, identify whether the lab was academic or industrial, would be considered quite ingenious indeed. Nor is the product of their activity describable in isolation. Both incorporate, as constitutive dimensions, the criteria, imperatives, and modes of judgment that, in one way or another, they had to take into account and which are not the result of personal initiative but refer to a collective practice. In other words, the constraints that need to be compared specifically designate a *practice* in the sense that this practice would not exist without the individuals who engage in it, but they are not intelligible independently of that practice, even though in some cases they contribute to its transformation. Therefore, it is the difference among *practices* that I intend to accentuate and, reciprocally, it is from the means that will allow me to articulate this difference that I infer the ecological identity and the possibilities of becoming of what I call "practice."

We can compare the problem presented by the difference among practices with the presentation of the "regimes of justification" proposed by Boltanski and Thévenot.[6] In both cases, practice or regime of justification, what is designated may be legible in an individual activity but what is read "situates" the individual with respect to a collective, relates her to what I have previously referred to as a "psychosocial type." And in both cases the inscription of the individual in the collectivity entails the problem of what one can legitimately claim or hope, the commitment that provides value or justification—"worth" (*grandeur*), as Boltanski and Thévenot write—for what she undertakes, the constraints she recognizes, whether she accepts them or transgresses them, the satisfactions she seeks. But I am not interested in a sociology of practitioners and scientific practices to the extent that they are "modern," and whose "worth" would then have as an ingredient their break with the order of appearance, of opinion or belief. I don't want to use this as a criterion for distinguishing scientific from technological-industrial practices, which have nothing specifically modern about them in the sense I have defined, for they freely acknowledge that fiction or opinion can be ingredients of their factishes. My project is one of experimenting with the possibilities of transforming this "economy of modern worth," of making those practices present that are likely to be of interest and which justify themselves on the basis of other criteria.

How is a modern practitioner likely to present herself to others? What does the expression "Rest assured, I'm an astronomer (physicist, molecular biologist, doctor, psychoanalyst)" mean if it is not coupled with "and not an astrologer (philosopher, vitalist, charlatan, hypnotist)"? How can we make it possible for a modern practitioner to present herself, justify her practice, draw attention to what interests her, without that interest coinciding with a disqualification? In other words, how can we "embody" in a mode of presentation and justification that is addressed to the "outside," and would thus be capable of intervening in the ecology of relationships among practices, what can already be read "from within," especially in the controversies through which the scope, rights, and obligations of a practice are discussed, challenged, and affirmed?

I want to provide an example of something that can be used to illustrate the specific constraints that define the "worth" of experimental factishes. In *The Invention of Modern Science,* I went so far as to write of a "new use of reason"—a grandiose expression if ever there was one. I claimed that this new use could be characterized as the invention of a new type of "fact," one that could be compared to the "neutral" fact that empiricist doctrines insist on invoking, a fact these doctrines characterize as "independent" of human interpretation and, as such, a solid foundation for the construction of "objective" knowledge.

The contrast I proposed between a raw, empirical, neutral fact and an experimental fact, a factishistic fabrication no matter how fully inhabited by human histories and, *as such,* capable of differentiating between the interpretations that are suggested for it, can be repeated in terms of *obligations.* A real "raw" fact, independent of us, like an earthquake or a tree falling on a passerby, is associated with no obligation involving the meaning it must be given: it is available for any interpretation, any creation of meaning, including the one that invokes some stoically endured accident. In contrast, the "experimental fact" reflects the singularity of the history in which it was produced. And the core of this history is that facts have value only if they can be recognized as being able to obligate practitioners to agree about their interpretation. And the practitioner who represents this fact, and claims to speak in its name, can do so only if she has first satisfied the strict obligations that will determine the value of what she proposes. The prospect of the controversy it will have to resist is constitutive of the fact and designates, as legitimate protagonists, those whom the fact claims it has the power to obligate.[7]

But that a fact effectively plays this role, that an experimental factish can be fabricated, can be expressed in terms other than those of obligations alone. There are *requirements* to be satisfied, which this time affect the "world" to which the practice is addressed. Experimental procedure requires that the phenomena to which it is addressed can be isolated and purified, that is, "mobilized," made capable of conferring the power of representing them on the one who questions them (by means of an experimental device). In other words, it requires the possibility of transforming a phenomenon into an "experimental fact," a *reliable witness,* capable of making a difference among those who interpret it.

To the mode of controversy that is constitutive of the "experimental fact" correspond, simultaneously and inseparably, both the requirement addressed to the phenomenon that it allow itself to be purified and mobilized, and the obligation, for the experimenter, to differentiate between two types of "artifacts." In one sense, any experimental fact is an artifact, a fact of art, a human invention. But the experimenter is obligated not to produce an artifact in a second sense, the sense that the "fact" could be recognized as a false witness, proving nothing, shown to have been created by an operation that claimed only to be a purification, or by an experimental device that influenced the results it should merely have made possible.

This first example is both general and specific. It is general in that it presents, on the one hand, the abstract character of the specific, "psychosocial" constraints I have referred to as requirements and obligations, and, on the other hand, the link between the satisfaction of those constraints and the "worth" of the event that celebrates the creation of each new type of experimental device, each new type of experimental factish. But the example is specific in the sense that it is a specific feature of experimental practice that requirements are entirely directed at phenomena, whereas obligations are entirely relative to colleagues and controversy. The "phenomenon," once it is defined as material for experimentation, is a source of no obligation, it is something that does or does not satisfy experimental requirements. Correlatively, there are no specific requirements that condition participation in experimental practice. Of course, the standard is that one must have the appropriate degree and be included in some authorized group. But nothing, in principle, excludes someone who is self-taught from inventing an experimental fact.

Requirement and obligation have the status of abstract constraints whose existence is not demonstrated by the ability to characterize them via some clearly determined mode of satisfaction but by the problem they pose, by the way they impose, on one mode or another, their relevance, whether through trickery or some more or less acrobatic negotiation. It was the error of normative epistemologists to think they could explicate, in the form of norms to be obeyed, the experimenter's obligations, as if the nature and limits of the authority the phenomenon confers were not in themselves what is at stake in discussions and controversies. But it was the error of contemporary relativists to deny, under the pretext that they were unable to supply them with a stable identity, that experimental research was singularized by requirements and obligations irreducible to just another argumentative strategy. Naturally, it is possible to show that, on a given occasion, a scientist has demonstrated his indifference to the distinction between "fact" and "artifact," or has simply ignored the ambiguity, and the multiple possible interpretations of what he claimed to be a reliable experimental witness, relying, for example, on his prestige, the strength of his rhetoric, or his connections with power. But what serves, in the eyes of the relativists, as an example to illustrate their thesis is not, for those of their readers who are experimenters, a lucid description of what they are all doing. It is an act of accusation that, if acknowledged, would be equivalent to condemning the scientist in question. Artifacts, operations of mobilization, are *not* all equivalent. Establishing their nonequivalence is a sign of the worth of an experimental practice.

Let's now examine how those two constraints, requirement and obligation, operate in a way that is detached from the specific case of experimental invention. We might be tempted to compare them to "rights" and "duties"—which are frequently coupled—but we should resist this temptation. Rights and duties refer to a problematic characterized by homogeneity and reciprocity. It is we humans, the subjects of Kantian practical reason, who can come to mutual agreement by respecting the rights of others and by recognizing the duties each of us has toward ourselves and others. There is no event here but an institution that should instigate acceptance, the loyalty of heart and mind. Requirements and obligations do not function in terms of reciprocity and, as constraints, what they help keep together is not a city of honest men and women but a heterogeneous collective of competent specialists, devices, arguments, and "material at risk," that is, phenomena whose interpretation is at stake.

Moreover, it is clear that requirement and obligation describe a kind of topology. We require something from something or someone. We are obligated by, or are obligated to—with, in some cases—the dimension of gratitude that the Portuguese language emphasizes.[8] The topology described distinguishes an "outside" and an "inside." However, it would be rash to use the example of experimentation to assimilate an "outside" or "material at risk" that may or may not satisfy certain requirements to the field of phenomena, and an "inside" to the specialists who investigate it. For each practice, it is on the basis of the definition of what is designated as "reality" and what will be asserted as "value" that the scope, implications, and problems of requirements and obligations can be specified. In other words, the practices whose topology is characterized by requirement and obligation are, by that alone, defined in constructivist fashion. They are not subject to any transcendence but, through the constraints that particularize them, bring about the nonequivalence that those transcendent agencies are most often responsible for establishing.

The distinction between requirement and obligation is crucial for the question of the justification or mode of presentation of so-called rational practices. The theme of "rationality" changes meaning depending on whether it refers to the register of requirement, where it is most often a vector of arrogance and infamy, or that of obligation, where it becomes synonymous with risk and challenge, not for opinion or ignorance but for the one who chooses to enroll in a practice that claims it. While the notion of requirement, taken in isolation, opens the door to relativist irony, that of obligation allows us to acknowledge the abuse of power constituted by relativist judgment, that is, the reduction it brings about of whatever engages the practitioner in forms of corporatist pretense, naive belief, even lies, which the relativist judge would be responsible for exposing. In constructivist terms, we could say that the production of obligations pertains to the register of creation, which *must* be acknowledged in its irreducible dimension, while the assertion of requirements presents the problem of the possible stability of that creation, of its scope, and of the meaning it proposes to embody for others. The concepts of requirement and obligation allow us to keep both the respectful ratification of claims to rationality and the relativist irony that judges them at a distance. They are available to "empirically" acknowledge what belongs to the order of an event and to "problematically" follow the actualization of the new situation that claims to be authorized by that event.

Naturally, the constraints found in requirements and obligations do not in themselves singularize scientific practices any more than they do the practices I refer to as modern. We can characterize the "worth" of technological-industrial creations in terms of requirements and obligations. We might even be tempted to extend this mode of characterization to the living. Every living being may be approached in terms of the question of the requirements on which not only its survival but also its activity depend, and which define its "milieu." And every living being brings into existence obligations that qualify what we refer to as its behavior: not all milieus or all behaviors are equal from the point of view of the living; and the difference is especially relevant when we address those obligations we impose on the living in the name of some knowledge we wish to obtain. Viewed in this generic sense, "requirement" reflects the normative and risky dimension of dependence on a milieu, that is, on what may or may not fulfill needs and demands. Any practice or living being depends simultaneously on diverse milieus. A physician depends on the answers of the sick body she addresses as well as on society, together with the series of institutions that make it possible for her to provide care. This requirement designates a set of interconnected modes of relationship, but the important point, reflecting its connection with the question of "worth," is that it always coincides with a principle of nonequivalence concerning what is defined as "exterior" or "milieu." The term "obligation" also reflects a principle of nonequivalence, but this time one that affects the "typical behaviors" or ways of proceeding of the practitioner herself, or the difference between that which, in her own practice and that of her colleagues, will excite, satisfy, disappoint, or be rejected as unacceptable. Obligation refers to the fact that a practice imposes upon its *participants* certain risks and challenges that create the value of their activity.

The constraints I have referred to as requirement and obligation do not in themselves specify scientific practices or modern practices, but the explication of those constraints can help us specify those practices in their creative dimension, bringing into existence both the reality to which they are directed, material at risk, and the values that define those risks. They may therefore be capable of following the singularity of the invention of so-called modern practices without sharing in the opposition between modern and nonmodern.[9]

5

Introductions

It is important that we not confuse the ecology of practices, as I've tried to describe it, with the practice of gardening. The gardener is free to select her plants, to arrange them as she pleases, to prune them as needed, and to try to get rid of whatever she considers weeds. She has the power to judge and to select. But at the opposite extreme, it is not a question of creating some ideal "vivarium," where different species are left on their own, some disappearing, others surviving, others proliferating. In fact, the practices of the gardener who selects or the vivarium creator who observes have little relevance for the question of an ecology of the practices, requirements, and obligations that would define it. The ability to select, as well as to avoid any interference, presupposes a radical temporal difference, a disconnect between the time of the human project and the time that characterizes the way beings affected by such projects relate to their milieu. Genetic engineering, which dreams of crossing this divide, of submitting living beings to the temporality of the human project, is the opposite of an ecology. Unlike the garden and the vivarium, the ecology of practices is defined first and foremost by the fact that the way those practices are introduced and justified, the way they define their requirements and obligations, the way they are described, the way they attract interest, the way they are accountable to others, are interdependent and belong to the same temporality. At that point, any argument, whether critical or condemnatory, justificatory or instigational, is a mode of intervention that *is added* to the interconnected ways in which the various protagonists involved address one another. But the possibility of an ecology of practices also requires the viability of a register of intervention that explicitly affirms the reciprocal capture that reflects and brings about any point of view recognized as pertinent. It requires what this text requires of its readers: abandonment of the opposition between "faithful description" and "fiction," between "fact" and "value," for an openly constructivist approach that affirms the possible, that actively resists the plausible and the probable targeted by approaches that claim to be neutral.

Not all "ecological" situations are equal. Without this statement—which does not conform to a subjective judgment in the sense that we "add" value to situations that are indifferent to the problem of values, nor to a project for discovering "the" value that would summarize all the others—the proposition of an ecology of practices would lack relevance, becoming a simple naturalizing metaphor for an appeal to a generalized goodwill. The argument implies a problem of creation rather than the problem of recognizing some basis that guarantees the difference between truth and illusion. That nothing is "natural" in nature or "naturally respectable" in society, that everything, from ecological situations to sociopolitical regimes and moral values, is in this sense an "artifact," has as its correlate the fact that I intend to create a new artifact. The obligation corresponding to this creation is to bring into existence, without recourse to any form of transcendence, in the very act of describing practices, as effect-artifact, a challenge and a problem that are detected at the same time as they are produced, and are found, as soon as they have been engendered, to have already been there "all the time."

I have emphasized from the outset that this challenge and this problem are here called "coherence." Therefore, I must try to "show," make perceptible—and not "demonstrate"—the possibility that by presenting and representing itself in terms of requirements and obligations a practice can affirm the existence, legitimacy, and interest of other practices with divergent requirements and obligations. This challenge, this problem, entails the creation of value and cannot be associated with the recognition of a value, especially in the name of Peace or the Good. Here, we may recall Gilles Deleuze's statement that Spinoza's Ethics is an ethology,[1] for "the opposition of values (Good-Evil) is supplanted by the qualitative difference of modes of existence (good-bad)."[2] Ethology, whenever human practices are involved, is based on productive, on performative experimentation with regard to modes of existence, ways of affecting and being affected, requiring and being obligated; and the substitution of judgment values by "ethological" values should not be argued for but produced or performed.

We should not, however, yield to the ease of untrammeled freedom. Every practice, including the one I am introducing here, may need to account not for the values it brings into existence and that enable it to exist, but for the coherence between those values and the means employed. Are the means I give myself, the approach to practices in terms of requirements and obligations, appropriate to the *problem* I want to bring into existence, that I wish to add to the problems that engage our different practices, namely, the escape from a generalized polemic that puts every practice in a position of disqualifying and/or in danger of being disqualified?

To be able to frame the problem requires, on the one hand, that the singularity of a practice can be described in such a way that the relationships this practice is capable of entertaining with other practices are not determined by that very act, and, on the other hand, that those relationships are not described as indifferent, subject to an organization we would be free to determine in the name of a shared ideal. Mutual agreement among practices cannot be decreed. It is not a matter of interdisciplinary "goodwill" or of the static distribution of the territories of each practice and the rules of noninterference. From this point of view, it is obvious that the means I have given myself are vulnerable to the most obvious misunderstandings: "requirements" and "obligations" could become a type of business card offered to each practice, or that a practice would offer itself, or all-purpose instruments for giving free rein to the widespread sociologizing temptation to catalog the variety of what one is dealing with. Would requirement and obligation then lose the abstract character of a problem—can I demand, am I obligated—and become a claim, or the attribution of a point of view?

Correlatively, we can ask how these concepts, which are intended to characterize the constraints that singularize each practice, situate those who would make use of them. Are these philosophical concepts, in the sense that the philosopher creates concepts not to refer to a state of affairs or a lived experience but to "set up an event that surveys the whole of the lived no less than every state of affairs"?[3] Or are they likely to contribute to the way in which a practice is able to present itself, in terms of the risk that engages it and not the disqualification of the other through which its own rights are affirmed? Or, wouldn't they specifically designate a practice belonging to the so-called social or political sciences, even the humanities, to the extent that these would recognize as their primary obligation the ability to address others as capable of becoming, and not as authorizing identification, which is to say, always, judgment?

To this question, there is no, there can be no, answer; for such an answer would presuppose the possibility of a judgment about what a philosopher, a physicist, a biologist, a doctor can become. But it is important to point out that, today, this basic impossibility is accompanied by a very specific difficulty. Let us assume, since I am proposing these concepts and, apparently, I am a philosopher, that they have something to do with philosophy. But the field known as philosophy has now become an asylum for what I would call "political refugees," researchers who have come there to ask questions that are directed—whether in a form that is critical, reflective, historical, or speculative—at a field of knowledge or a specific practice, but that cannot be asked within that field. Originally, I was just such a refugee, and it was also, and almost by accident, that I was able to experience what it is that philosophical concepts appeal to, in the sense that "the concept belongs to philosophy and only to philosophy."[4] This experience did not directly answer the questions that had led me to give up chemistry, but it gave them greater acuity, prevented them from closing up too quickly, from being satisfied with historical, epistemological, or critical solutions. I turned away from the question of knowing what so-called scientific practices are to embrace the question of what they might become.

If there is something singular about philosophy, as understood by my experience of it, it is that it no longer amounts to anything if it merely refers to what it brings into existence as having the power to confirm it, or if it refers to itself as having the power to prove what follows from the problem it creates. In other words, if the sciences can maintain, at their own risk, a relationship of collusion with power, this relationship constitutes a primordial risk for any philosophical problem and, in this sense, there certainly exists a singular affinity between philosophy and the question of becoming. But the primordial nature of this risk does not mean that philosophy alone embodies the question of becoming, even less so since other practices can refer this question to philosophy. Rather, it means that it is only when those other practices are involved in relationships capable of actualizing the challenge of the question of becoming that philosophical practice can become "one" practice among many, adding itself to the others without the slightest temptation to regulate them, or appropriate their risks, or judge them on the basis of the relationships of collusion with power that mark them. The challenge of becoming creates the possibility of the coexistence of distinct problems correlated by the way in which, to borrow an expression from Deleuze and Guattari, they *allude* to one another.[5]

The notions of requirement and obligation are certainly capable of functioning as philosophical concepts, but that does not mean that they constitute an ecology of practices in philosophical practice or introduce some form of "philosophical reflection" within the fields they problematize. Their use is, first and foremost, delocalized, identifying a priori neither the user nor the field of use. They function as operators intended to make perceptible, in the very way they must be reworked to earn their relevance in each practical field, the topological transformations that mark the transition from one field to another, which is to say, the qualitative differences between the respective types of events that can be celebrated in terms of a satisfied requirement or an obligation fulfilled.

The possibility of constructing "delocalized" concepts, which guarantee the ability to travel anywhere and to be at home wherever one happens to be, has always fascinated Plato's heirs, and has always made them vulnerable to the seductions of power whenever they judge states of affairs in the universal terms such concepts claim to authorize. Doesn't this demonstrate that those concepts have achieved universality, that is, that they authorize a judgment from a fixed point from which local, changing, and, therefore, deceptive appearances lose all but anecdotal interest? The requirement to be able to rediscover the same, here or elsewhere, the same "man," the same moral law, the same divisions between truth and fiction, between nature and culture, would be nothing more than an innocent obsession if the statements it engenders didn't have the terrible ability to become code words, ready-made, migrating unimpeded from the philosophical search for fundamentals to the claim of the right to assert the universal anywhere and everywhere. What has transpired between the philosopher's meditation on human freedom and the declaration of a Belgian politician: "We are opposed to any form of dependence, mental or physical"—and, therefore, opposed to the legalization of marijuana in any form?[6]

Obviously, I have a very different kind of delocalization in mind: to bring into existence the experience of here *and* there, the experience of a here that, by its very topology, affirms the existence of a there, and affirms it in a way that excludes any nostalgia for the possibility of erasing differences, of creating an all-purpose experience.[7] My hypothesis is that the concepts of requirement and obligation can serve as vectors for this experience to the extent that they shift attention from visions of the world and the great metaphysical questions that claim to be valid for each and every one of us, to the singularity of what "matters" here and not somewhere else, to "technical details" that no one but the practitioners involved would deem worthy of interest but which, for them, are the difference between "worth" and failure. In this case, delocalization does not signify the possibility of avoiding disorientation, of discovering, no matter where we are, behind the anecdote, the same categories of universal judgment; nor does it signify a "taste for the exotic," the search for strong emotions for which somewhere else is simply a means. It implies a culture of disorientation, whose touchstone is not so much an openness to others but the ability to "introduce oneself," the condition for any civilized encounter.

The problematic of such an encounter defined the situation in which I learned the necessity of this "culture of disorientation" as an antidote to judgments that suppress any possibility of encounter, a situation in which the theme of an ecology of practices and constraints, which I now call requirements and obligations, started making sense for me. I began to discover what it means to "become a philosopher of science" within the Department of Physical Chemistry at the University of Brussels, under the direction of Ilya Prigogine. Working as a doctoral student in philosophy, I discovered the importance of having the researchers gathered there forget that I was a philosopher—a far too exotic label that was worthwhile only as a joke or for those "unanswerable" questions that had relevance for none of my colleagues. And, in the end, this suited me perfectly to the extent that I myself didn't know what that label demanded or implied. However, I discovered my obligation in my encounter with what, for Prigogine, was the true great work, compared to which the exploration of nonequilibrium physics, for which he went on to win the Nobel Prize, was only a first step. Prigogine was attempting to incorporate, within the "great laws of physics," those that authorize the "vision of the physical world" to which, as we saw in the case of Max Planck, the faith of the physicist is addressed, an expression of the difference between "before" and "after," which physicists refer to as "time's arrow." And this attempt, whose ambitiousness I learned to appreciate, presented itself in a way that was strange for a philosopher, but apparently ratified by the majority of physicists, whether or not they were sympathetic to Prigogine's approach: it was indeed up to time's arrow, with Prigogine as its spokesman, to demonstrate that it *had* to be taken into account in a physical world that, until now, seemed quite capable of ignoring it. Absent this, physicists would go on teaching that the difference between past and future is merely a question of probabilities, which were themselves related to the approximate character of our measurements and manipulations. For the being who would observe and manipulate the world in the ideal way that would give to the laws of physics their full power, this difference would have no meaning.

A strange situation. Even as an apprentice philosopher, I

knew enough about the "tricks" of my trade to laugh. The retaliatory argument, which traps the speaker by demonstrating that what she is saying is contradicted by the fact that she's saying it, was sufficient: if you want to persuade us that freedom doesn't exist, why take the trouble to do so unless you believe, in spite of everything, that we have the freedom to recognize the strength of an argument? In this case, it was sufficient, more than sufficient, to recall that the well-known laws of physics that affirmed the equivalence of "before" and "after" were only made possible—and let's not even mention human history and the practice of physicists—by operations of measurement, and that even the lowliest measuring instrument refutes this equivalence. In one way or another, these laws affirm a world in which the operation of measurement would be meaningless and it would thus be impossible to state them. I was thus free to scoff that one had to be a physicist to assign them such authority that one could, even for a moment, consider denying, in their name, what they presuppose and what every thinking and speaking being presupposes—including the physicist who announces to a shocked public that the difference between past and future is relative only to the "macroscopic character of our measurements." For the physicist hopes to create a difference of a vastly different kind than "merely macroscopic" for those in the audience.

Indeed, "one had to be a physicist. . . ." Had the comment assumed an edge of irony, the encounter would have been impossible. Or else it meant that I was dealing with a "somewhere else" whose singularity my philosopher's "here" might allow me to catch sight of. This alternative excludes a third scenario, however: that I accept the position of the "dumbfounded public" that forgets what it knows and attribute to the physicist the power of voicing an (objective) truth, a truth that transcends the practices that produced it.

Once the principle of the encounter was accepted, it situated me.[8] Because philosophy requires that I not follow the bearers of truth, I had to maintain my "here," resist the conversion that would have made me acknowledge that Prigogine's quest was decisive for everyone, that the question of the relationship between time's arrow and the laws of physics was one of those "great" questions that by rights must interest all human beings. It was a resistance to what might be called "scientism," which identifies the sciences as the "guiding brain" of humanity. But I also discovered that, to develop an interest in Prigogine's work, to make it interesting in a way that was not scientistic, it wasn't his "requirements" that I later had to understand so much as his "obligations."

Prigogine's requirements involved the physical–mathematical beings that he and his colleagues worked with. A specialist's competence would have been needed to be able not to follow a posteriori but to accompany the process of concocting the beings the mathematical physicist addresses as if they had an autonomous existence, testing their properties in order to determine whether or not they satisfied his requirements. Requirements that those beings had to satisfy in order to earn, in this case, the status of representative inhabitants of the new "physical world" that was at stake, the one that would coherently extend the mode of intelligibility associated with classical and quantum laws while affirming the difference between "before" and "after" that those laws deny. On the other hand, it was through the obligations entailed by the tests those beings had to undergo, the dreams, fears, doubts, and hopes that led to their invention, that, as a philosopher, I discovered the singularity of this practice. And, most importantly, I learned how little the obligations of the philosopher, her dreams and fears, her doubts and hopes, mattered to the physicist, and how much the reworking of his connection to his own tradition, the continuation of an increasingly intimate dialogue with his forebears—Hamilton, Poincaré, Boltzmann, Bohr, Einstein. The physicist's dreams are not nourished by the risks of philosophy but by those that engender the values of his own tradition.[9] The physicist does not "lack" philosophy, but it may be the philosopher's job to forge the words and phrases that can be used to "transmit" those values to others without confusing them with a commitment to a truth, whole and naked, in which all of humanity would recognize itself.

But the philosopher was also there as a political refugee, coming from another landscape of practice, chemistry, a landscape swept by the contrast between experimental invention and submission to a hierarchy that reduced chemistry to a particular application of the universal laws of physics.[10] It may also be that the unfinished chemist and the apprentice philosopher both experienced this "culture of disorientation" that is now, for me, a humorous synonym for truth.[11]

The chemist, then, is never very far whenever the philosopher questions herself about the history of physics. If the making of the first true experimental "factish," the Galilean ball rolling down an inclined plane, hadn't intervened in a history where the unification of the celestial world with terrestrial nature was at stake, maybe the values of the tradition to which Prigogine belonged would have been different, not associated with the strange metaphysical passion characterized by the continuously renewed idea of the "laws of nature." Perhaps the event that constitutes the creation of every new experimental factish would have been celebrated as such, rather than "dismembered," framed, according to some, within a hierarchical order dominated by reference to universal laws for which the new factish would offer a particular kind of expression, and, according to others, exemplifying the fulfillment of the general, "neutral," and all-purpose obligations known as "methodology." The "here" I encountered in Prigogine's ambition to incorporate the arrow of time within the very framework of the "great laws of physics" produced a connection with a possible, speculative "elsewhere" where the requirements and obligations of experimental invention would be celebrated as such, the "elsewhere" of a history in which the laboratory and experimental proof would have been associated with the continued creation of new beings, transcending the practices that caused them to exist. It is not with Prigogine but because of Prigogine that I came to the conviction that it was important to celebrate experimental success, the successful differentiation between "fact" and artifact to which experimental proof obligates us, not to authorize a vision of the world, but rather to create beings whose autonomy is *specified* by the requirements that the obligations of proof have brought to bear upon them, requirements they were able to satisfy. Whether we direct our inquiry to "phenomenological" physics—whose laws do not refer to "nature" but to a reality staged in a laboratory—chemistry, Pasteurian microbiology, or molecular biology, we can say that experimentation poses the problem of a reciprocal capture whose ideal is to be equated with *unilateral* capture: the act of experimental creation that was recognized as mere purification can, by that very success, claim to disappear, claim to have simply "made available" to knowledge a reality that preexisted it, the way oxygen preexisted the aerobic metabolism that captured it.[12]

Little by little, I learned that I had to direct my inquiry to the distinction between dreams and fears, doubts and hopes, between what was celebrated as an obligation fulfilled and a requirement satisfied, if I was to approach those I wished to encounter. To approach biologists as a philosopher meant asking myself what difference it made to them that their practice addressed problems it didn't invent because, in one way or another, living beings were already introducing a "solution," were endowed with relative autonomy. What if the practical question posed by this autonomy of living beings were expressed in terms of specific obligations? What if the scientist was no longer obligated by her invention alone and by the values she brings into existence "here"? What if she were dealing with something "else," an "elsewhere" she was not free to define in terms of her requirements alone?

What might become central to bringing this tension between "here" and "elsewhere" into existence in place of the ongoing confrontation between the fratricidal twins of "reductionism" (the extension, here and elsewhere, of the same explanatory strategies) and "holism" (the creation of barriers protecting the "autonomous whole" against those strategies) is the price paid by those affected by the knowledge we construct. And this should not be understood in the sense that knowledge in this case would be destructive or mutilating by definition. Rather, the primary obligation of the scientist would be to avoid dreaming the dream of experimental creation, of turning a living being into a reliable witness for its own functioning, that is, of defining it through the unilateral requirements of proof. Such an obligation, however, should not be seen as a loss. It celebrates that which gives its singularity to an approach in which knowledge must be referred to as "encounter" and "learning" because it is addressed to a being that itself presupposes and requires a milieu. This does not preclude experiments but demands that experimenters try to make explicit what a being presupposes and requires in order to formulate the questions they may require it to answer. To question the interactions and challenges to which a being lends itself, and ask oneself what that being "makes" or "has made" of what it requires or has required, implies an appetite unlike that of a creator. The experimenter is a creator. She brings into existence a being that will serve as a reliable witness to whatever determines that being's behavior. In this case, the appetite is that of an investigator. The investigator defines her terrain, that is to say, the elements that, in one way or another, were likely to, or had to, play a role in the matter requiring clarification. But she cannot assume that those elements will explain the matter. She must reconstruct the intrigue, which is always one particular intrigue among the variety of intrigues compatible with that same terrain.

The difference between "here" and "elsewhere" is not a matter of irreducible opposition, an insurmountable limit or uncrossable border. "What is it capable of?" is a question that can inhabit the dreams of the experimental scientist as well as the biologist. But, depending on the situation, this question can define what it is addressed to in very different ways: to a being that is supposed to "obey," that is, supposed to be available to the manipulation that will enact this question, or to a being whose mode of existence depends on the way in which it has already, elsewhere, answered the question.

If the ecology of practices is intended to make present in their singularity the requirements and obligations of different knowledge-producing practices, it can also question certain practices because of what they require. This is what I would now like to demonstrate with respect to the setting invented by psychoanalysis.

Apparently, the requirements of psychoanalysis affect the practitioner. They designate the analysis that has made him capable of an "analytic" relationship with the patient (or the Lacanian analysand). As has so often been remarked since Freud, the only analytical instrument is the practitioner's ear, and the way he listens is the business of analysts alone. Correlatively, only a genuine analyst is qualified to authenticate the practice of another analyst. No questioning on the part of the "patient" is acceptable as such, or, more accurately, none will be accepted other than as a symptom, the real raw material of the cure. Consequently, analysts seem to constitute a radical "elsewhere," a closed community. Although no one has had a problem with the fact that Galileo was not, and with good reason, trained in an experimental laboratory, the fact that Freud, the first analyst, is also the only one never to have been analyzed by anyone, is a perpetual subject of concern for psychoanalysts.

Opponents of psychoanalysis have often emphasized the

fact that it has never submitted to the obligations of proof, that the cases it relates could not in any sense serve as "reliable witnesses" of the categories it invokes in presenting them. To the extent that the unconscious, intrapsychic conflicts, and resistance could be presented as factishes capable of legitimizing the practice of analysis, capable of confirming that the requirements that define the practice are indeed those that must be satisfied by anyone who addresses mental suffering, those criticisms are legitimate. And in that case, the closed nature of the analytic community turns psychoanalysis into a genuine instrument of war, not a practice whose obligations would express singular risks and values.[13] Psychoanalysis, whenever it claims the power of proof, defines all humans as "obligated" by its practice. For its practice could then be said to concern "anyone," because anyone should seek the truth of her suffering rather than escape it through the multiple "rationalizations" that prevail, here as elsewhere, among the modern builders of factishes just as they do among nonmodern adepts of fetishes. And the analyst or his allies would then be capable of judging anyone depending on their understanding of this, depending on whether or not they recognize that not all methods of addressing their suffering and their symptoms are equally valid.

"What do you want of us?" "How do you define us, you who claim to speak in our name?" To ask such questions of psychoanalysis is, as is always the case with the ecology of practices, to produce an active proposition, one that forces it to sort through its claims. Yet, unlike propositions addressed to the producers of knowledge who claim they create a difference between a scientific statement and fiction, here the question is also about the claims of psychoanalysis to constitute a "modern" practice focused on the creation of a difference that tells us who—as "moderns"—we are. In this case, the ecological proposition engages the question of modernity's unknowns.

6

The Question of Unknowns

In the foreword to *Difference and Repetition,* Gilles Deleuze writes: "The time is coming when it will hardly be possible to write a book of philosophy as it has been done for so long."[1] In my case, what is no longer possible, the thing whose impossibility has created the conditions for this book, is the forgetting of the "unknowns"[2] in the question I want to ask: forgetting what is concealed by the familiar landscape in which the philosopher walks, a philosopher confident she will meet only her kin, and not all of them close relatives. They may be a bit remote, irritable or skeptical on occasion, as arrogant as the newly rich from time to time, but they are always likely to share a common history, though it is one of family quarrels, arguments about inheritance, and bad marriages.

For a philosopher, to ask the question about so-called modern practices means, regardless of whatever original variations she might introduce, returning to a traditional question. It is normal, and was predictable, that philosophy would not stop inventing the means to question and to understand practices that seem to create other answers to its own questions, other solutions to its own problems. It is equally normal and predictable that the question of physics had to be a starting point in this research, for isn't it a source of passionate involvement and productive of statements resembling those of metaphysics? Finally, it is normal and predictable that this research should question the strange terrain it interrogates: a terrain that has been occupied, identified, and marked by the rules of good conduct a thousand times over. And yet it remains strangely wild, divided as it is among legitimate occupants, and continues to be intersected by startling pathways that appear, for better or worse, to ignore boundaries and rites of passage. It is a terrain that has been purified a thousand times over by all sorts of sacrificial deaths, and where a thousand and one monsters and hybrids still proliferate on the bodies of the sacrificed dead. Perhaps it is slightly less normal, a hair less predictable, that while using a classical approach, a philosopher would be willing to accept the challenge of adopting a deliberately uncritical standpoint, of exploring a terrain actively stripped of what allows philosophy to judge and disqualify. In short, the challenge of "thinking with" unknowns.

However, failure to forget the "unknowns" found on this terrain implies a failure to forget how philosophy itself is a participant and how, in particular, it is an ingredient whenever, in questioning modern practices, it accepts, implicitly or explicitly, the fatal nature of the destruction of traditional practices presupposed by the exclusion of the sophists. I've emphasized the fact that the excluded sophist is defined with reference to a malleable and docile opinion, which lacks a fixed point, rather than references through which traditional practices might "fix" their operations, which we denounce as fetishes. The use of the *pharmakon* attributed to the sophist appears to allow us to "understand" such practices, just as, today, "suggestion" allows many to claim to "understand" nonmodern "fetishistic" practices. In both cases, such a characterization appears to arise from a healthy use of critical lucidity, but it primarily reflects a disqualification that very clearly authorizes the exclusion of practices so characterized. It was to re-create the hesitation in such judgment and the sense of kinship it establishes among all those who endorse it that I used the term "factish" to refer to the neutrino and the other existents that, because of modern practices, are now official ingredients in our history. The question of the ecology of our practices, for which I used the figure of the "nonrelativist sophist," does not determine its own limits. It would be nothing more than a contract among scoundrels in a field of corpses if it wasn't haunted by the presence of what it must claim to be unable to define—we don't know what a fetish is.

With the problem posed by psychoanalysis, the question of unknowns assumes clarity. Unlike certain practices that I consider grotesque, if not literally obscene, and whose rejection, in my case, is self-evident, a practice such as that of psychoanalysis puts me in a *critical* position.[3] If the ecology of practices is not, as I have already noted, the practice of a gardener who selects, sorts, and eliminates, if it produces, rather, "active" propositions that offer practitioners the possibility of presenting themselves in a "here" that resonates with the "elsewhere" of other practices, how is it that, in this case, the proposition metamorphoses into a challenge, a challenge that concerns what may be the crucial claim of psychoanalysis? For, to "propose" that psychoanalysis abandon its title as a "modern practice" has little to do with "proposing" to a physicist that she disconnect the power of factishes from visions of the world that seem to authorize those factishes. The event that constitutes the invention of physics does not have as a crucial component the claim that it constructs a "vision of the world," whereas the invention of psychoanalysis would be unintelligible without the claim that it introduced psychotherapy into the landscape of modern practices.

The road would be dangerously rapid, here, leading from challenge to denunciation. The sciences, whenever involved in human experience, would be transformed into "objectivizing" and, therefore, destructive endeavors. "Moral" values, those of tolerance or postmodern relativism, would have to take over whenever it was a question of human behavior, and ensure mutual recognition among groups and individuals. And these are the values that would now nurture psychotherapy.

To express myself more directly, I would say that the only truly tolerant and relativist undertaking that I know of is capitalism. It alone is capable of radically aligning disparate practices and values only to turn against those whose destruction would be of interest to it; for it is radically indifferent to whatever binds them and is itself bound by nothing, even its own axioms of the moment, which have nothing at all do with requirements or obligations.[4]

Moreover, as with the general, neutral reference to what is called modern rationality, references to the equally modern tolerance for "cultural" practices or "properly human values" that would need to be defended from "science," for me serve as alarm signals, for those references dramatize the limitless nature of a rationality that could, by right, but should not, destroy what gives meaning to human life.

At the end of his life, Freud recognized the impossibility of distinguishing psychoanalysis from other psychotherapeutic practices based on their respective therapeutic efficacy. He wrote that it was, like education and government, "an impossible profession."[5] There is nothing innocent about these three "professions." Government, education, and medicine[6] refer to the threefold practical field by which every human society invents itself, that is, the way in which the relationship between the individual and the collectivity is constructed in all societies, modern or not. However, Freud doesn't bow down before the difficulty of these "professions," he celebrates the grandeur of a rational modernity that has destroyed their "possibility." Tolerance is not a virtue cultivated by psychoanalysis, even though Freud acknowledged that a certain amount of "suggestion" could be alloyed with the pure gold of analysis. Psychoanalysis must be the vector of the "drama" of modernity, and it is great because it alerts us to the risk of destroying not only the "nonmoderns," whose fate is sealed, but possibly ourselves as well, given that what it consigned to the world of illusion and "fetishism" was, in effect, a condition of possibility for the three, now impossible, "professions."[7]

The tolerance that claims to protect that which, in itself, is doomed, or the heroic affirmation that arises from the fact that what can be destroyed by antifetishistic effort must be destroyed, regardless of the price, are two sides of the same coin. And it is when this coin is taken to be valid currency, when a practice confirms the insuperable dilemma between uncertainty and destruction, that the ecological proposition changes its nature and becomes confrontational. I refer to all the practices that impose this substantive change as "modernist." They are what gives resonance to the question of unknowns, this "elsewhere" that they deny. Although the scientific factish makes its appearance accompanied by those for whom it makes a difference, those whose "competence" is reflected in the fact that they are kept in suspense by the relation they create and which creates them, the references created by modernist practices are supposed to make a difference for "humanity" as such. This is obviously reflected by a change in the nature of obligations, which, in every scientific practice, affect its practitioners. Modernist obligations produce the machinery of warfare and the coded language that enable any number of institutions to operate in good conscience because they can attribute the violence of their effects to the price legitimately demanded by any truth that challenges human illusions.

How can the ecology of practices be extended when its propositions lead to confrontation? Is it possible to characterize practices that are bound by their "modernist obligations" to the power to disqualify without betraying the immanent characterization of our practices in terms of requirements and obligations, that is, without judging modernist practices in terms of some form of transcendence?

Here, it is vital to recall that Freud's "three impossible professions," governing, educating, and healing, refer above all to *technical* practices and modes of thought: in their case it is not a question of proving or (re)constructing a story but of "fabricating." Any number of defenders of "uniquely human values" bow down to "science" but question "technology." That is why the term "technoscience" does not reflect the interweaving of scientific and techno-industrial developments but announces the radicalization of a critical position that the distinction between a "disinterested science" and a "dominating technology" can no longer silence. Here, I am attempting the opposite. What if such practices—apparently bound to the code words of modernism—were to recognize that their requirements and obligations are about fabrication, that is, are required to ignore the distinction between scientific claims and fictions? Would such recognition enable them to invent a mode of presence that causes the unknowns in the modern question to resonate? In line with this speculative proposition, it is the requirements, constraints, and obligations of different types of creative, transformative and fabricating relationships that would need to be made present. Thus, psychoanalysis, rather than seeing itself as a "modern technique," authorized by mental suffering and struggling with the fictions that silence what such suffering requires, might define itself as the *art of influence,* an influence that reflects neither truth, in the sense that such truth could be demonstrated, nor the arbitrary power of suggestion usually opposed to truth. What psychoanalysis seems to require of its practitioners would then be defined as closure, thereby defining a technical obligation; and what psychoanalysis required of its patients would become the condition required by its practice—what must be accepted for this technique to work.

Closure and condition are what usually allow us to disqualify nonmodern techniques, to denounce their initiatory nature and the way they require those who address them to acknowledge that they are dealing with an "elsewhere" where the standard rules don't apply. But maybe this elsewhere refers to the unknowns in the modern question. In that case, the space in which factishes and fetishes respond to one another could be delineated by contrast. We could speak of a "fetish" wherever obligations no longer appear as a means of verifying that requirements have been successfully satisfied, as is the case with our factishes, but bring about closure and thereby create the primordial constraint. Requirements, however, would be associated with what can only be satisfied performatively, through the implementation of technique, and relative to the constraint that technical obligations cause to exist.

Before allowing our scientific "factishes" to enter history, we demand that they be able to affirm that they existed "before" the practice that brought them into existence, which practice could then be explained through those factishes, in terms of "discovery." This demand need not be criticized, it is part of the conditions of existence of these modern beings, but its implications have to be evaluated. In one way or another, the creators of factishes depend on the stability of what they are addressing, a stability that allows mobilization to be presented as mere staging.[8] Whenever this stability is no longer dependable, the technique that mobilizes a reference can, or should, no longer do so in terms of the claim characteristic of so-called modern technique: to "obey nature in order to control it." Reference is no longer defined in terms of requirements—this reference should have the power to demonstrate its autonomy—but in terms of obligations—the mobilization of this reference should entail obligations that bring technology and the technician into existence.

The immanent mode of existence of practitioners who could be called fetishists—and the Freudian unconscious would then indeed be a fetish—would thus require the presence among us of beings capable of bringing about obligations in a way that blurs all opposition between "truth" and "fiction." But how can we then ensure that the question of the unknowns of the ecological problem will lead us to what every modern Western philosopher has learned to consider a part of the past, namely, the obligations attached to the "sacred"? Isn't the ecology of practices closing in on what we customarily call a "religion" or "cult"?

"Where angels fear to tread" is how Gregory Bateson titled the book he was working on at the time of his death.[9] "As I write this book, I find myself still between the Scylla of established materialism, with its quantitative thinking, applied science, and 'controlled' experiments on one side, and the Charybdis of romantic supernaturalism on the other. My task is to explore whether there is a sane and valid place for religion somewhere between these two nightmares of nonsense. Whether, if neither muddleheadedness nor hypocrisy is necessary to religion, there might be found in knowledge and in art the basis to support an affirmation of the sacred that would celebrate natural unity. Would such a religion offer a new kind of unity? And could it breed a new and badly needed humility?"[10]

Why should we be surprised that we might fear to venture forth into what Bateson calls the "epistemology of the sacred"? The "sacred" refers to what, from the modern point of view, is characterized as something that is not subject to any "consistent" knowledge, the kind epistemology might establish or speak about. "No one is truly modern who does not keep God from interfering with Natural Law as well as with the laws of the Republic. God becomes the crossed-out God of metaphysics, as different from the premodern God of the Christians as the Nature constructed in the laboratory is from the ancient *phusis* or the Society invented by sociologists from the old anthropological collective and its crowds of nonhumans."[11]

I might be tempted to summarize even further: couldn't an "ecology of practices" provide a more economical answer to

Bateson's question than his own "epistemology of the sacred"? It may indeed serve as an antidote to the Scylla of "established materialism," recognizing this "materialism" to be, as Latour claims, one of those lazy labels that "serve only to dissimulate the work of forces and make an anthropology of the here and now impossible."[12] But what about Bateson's Charybdis, "romantic supernaturalism"? What about the passion of conversion that has overloaded worthwhile questions like the Gaia hypothesis? What will be the antidote that recognizes the problem? And how are we going to avoid our tradition's preferred vice: constructing a convenient argument that has, as if by accident, the power to dissimulate or silence a question it feels uncomfortable addressing? Bateson's question gives rise to an unknown that, although it does not yet have the ability to be "considered," must mark the way we present the present.

I have named this unknown "cosmopolitics." Within the tradition of philosophy the term is Kantian in origin. The *jus cosmopoliticum* is associated by Kant with the project for "perpetual peace" corresponding to a "natural destination of mankind," in the sense of an idea that demands to be constantly pursued rather than a constitutive principle that would turn this destination into an object of knowledge.[13] The possible unification of all people through certain universal laws involving their possible commerce was for Kant a perspective not devoid of plausibility. In fact, the progress toward this unification was already manifest for him in the way the violation of law anywhere on Earth was beginning to be resented by his contemporaries. Today, we have reason to complicate this point of view. Although the idea of peace among various peoples must have some significance, we need to start not like Kant from promises the West might flatter itself for propagating, but from the price others have paid for this self-definition. It is not so much peace that we have brought to other peoples and ourselves, but a new scope, new methods, and new modalities of warfare.

It is, therefore, in contrast to Kant, rather than as a follower, that I borrow the term "cosmopolitics," and this contrast finds its initial expression in the constructivist approach of so-called rational practices. Methodological law, the Kantian "tribunal" capable of examining such practices from the point of view of the rules with which they must conform, is, and can only be, a weapon of war against everything that appears to infringe those rules. The neutrino—fortunately—does not owe its existence to a rule-abiding practice, and its coming into existence violated, and caused the restatement of, many rules concerning the definition of the legitimate mode of existence of a physical particle. To claim that the neutrino's mode of existence is that of a "factish," product and producer of a practice, existing through it and causing it to exist, then serves as a first step in exiting the Kantian perspective, where peace has to be "our" peace, where commerce is limited to goods and ideas, to the detriment of the multiple worlds brought about by our factishes and our fetishes. This doesn't simplify the problem but disconnects the question of the sacred from that of conversion. If there is a "sacred," even in the neutrino, that sacred is incapable of demanding the conversion of anyone not involved in the laboratories that require its reference. Therefore, if there is to be "religion" in Bateson's sense, the unity it celebrates will not be the one produced by an entity finally recognized as having the power of assembly. It is in terms of obligations rather than requirements that the unity of here and elsewhere can be asserted, the copresence of that which, at the same time, claims to be heterogeneous.

The Batesonian question of an "epistemology of the sacred," the unknown that our heterogeneous factishes and fetishes cause to resonate, contradicts the unifying references that, in the name of peace, demand conversion. Like the cosmopolitical question itself, it is directed at the modern tradition, at its contempt for fetishes, its fear of the *pharmakon*. The unknown that the question of a "(practical) epistemology of the sacred" causes to resonate is to determine whether such an epistemology could sustain the obligation to resist the code words that transform the singular, impassioned adventure of this very particular tradition into neutral "basic statements," secular statements that must be endorsed by those whom this epistemology has disqualified. Whether or not it could serve as an antidote for this contempt and obsession. Whether or not it could contribute to the creation of metastable regimes for articulating our practices, other than the one in which predatory relationships prevail.

"If cruise missiles reach me in the vineyards as I leave the house, I don't want to have to kneel down to 'reason' or a 'delinquent physics' or the 'madness of mankind' or the 'cruelty of god' or 'Realpolitik.' I don't want to invoke any of these complicated explanations, which confuse with power the reason why I am being killed with facts. In the few seconds that separate illumination from irradiation, I want to be as agnostic as someone can be who is witnessing the end of the old Enlightenment, and sufficiently confident in the divine and in knowledge to risk waiting for the new enlightenment. They won't get me. I won't believe in the 'sciences' *before,* and I won't despair of knowledge *after,* a force ratio assembled by certain laboratories explodes over France. Neither belief, nor despair. Agnostic, I told you, and as agnostic as one can be."[14]

"They won't get me!" The problem of unknowns with which I want to open the question revolves around the meaning assigned to this heartfelt cry. In its most commonly accepted meaning, which is not Latour's, this cry is uttered by someone who intends to remain alive in spite of the dead ends, impossibilities, and paradoxes our tradition leads to and celebrates with a certain pride.[15] It is the grand Platonic theme of leaving the cave that has been reinvented according to the new terms of heroic asceticism rather than the old promise of progress. We should be able to accept that progress no longer refers to a stable direction beyond multiple and confused appearances; that its mark is the very wound it engenders, the farewell (pronounced by some in the name of all) to a lost sense of security.

For a human to be marked by this wound, she must have first agreed, body and soul, to the obligations that are both confirmed and defined as impossible to fulfill by that wound. The expression "they won't get me" may then change its meaning, if it asserts that "they won't get me with the cave trick again." For this is the great ploy of this tradition, its claim that it, and it alone, has discovered the path distancing humans from idols and fetishes, from everything through which the "others" were "got" or possessed. In this way, not getting "got" becomes a way of ensuring that the loss of security actively diverges from the theme of the wound of truth, a way of being able to state that this wound itself is the mark showing that we have indeed been, and we continue to be, "got."

"They won't get me" then serves as a challenge to resist what has "gotten" us, namely, the belief in the power of proofs capable of disqualifying what they have no means to create, that is, because they are unable to transform what they disqualify into a reliable witness able to impose the way its testimony should be interpreted. Agnostic, yes, but actively agnostic, is the cosmopolitical question, the question that gives resonance to the unknown implied in what we have been able to create, and prevents what we have been able to create from obscuring its own conditions.

"Where angels fear to tread," wrote Bateson. This is not a limit, an abandonment, an abdication of reason but the constraint that may lead to a resumption of the Kantian idea of a possible commerce among the peoples of the Earth, through the deliberate and actively agnostic negation of the universal laws before which that commerce is supposed to bow down in order to ensure its pacific nature. Commerce, whenever it is confused with the now well-known ideal of the free circulation of goods, is nothing but a generalized state of war, the determined destruction of anything that hampers circulation and blocks the universal law of exchange. As for the practice of commerce, in the ecological sense of the term, it can suspend the certitudes of warfare but it cannot cancel the risk of war. The challenge of the cosmopolitical unknown is to give resonance to the obligations of such practice, to prevent it from disappearing into the norm of equivalence ratios, and the force ratios that such equivalence always reflects. But to construct the question in a way that avoids conversion, we must begin by reinventing questions wherever we have been converted to believing in the power of answers.

BOOK II

The Invention of Mechanics

POWER AND REASON

7

The Power of Physical Laws

Physics, today, is haunted by laws, and as long as this is so, as long as it presents itself as the science that discovered that nature obeys laws, it will stand as an obstacle to any ecology of practices. Like Max Planck at the beginning of the twentieth century, and Steven Weinberg and many others today,[1] the physicist will continue to claim to have discovered laws whose objectivity is in itself a denial of ecological thought, laws that, although with different notations and in different languages, would also have been "discovered" by intelligent extraterrestrials.

This wouldn't be such a problem if physics didn't serve as such a bad example. In the following pages, I examine two cases, chosen because they illustrate this "bad example" in two different ways: the mind–body problem and the mathematical theories of economic markets. In the first case, it is the power of physicalist explanation that fascinates us; in the second, it is the power of the highly abstract mode of representing the dynamic state, apparently detached from any particular physical model, that is of interest.

How can we characterize physics's role in all this? I first want to point out that, here, there is no question of invention or, therefore, of an event. I'll return to the physical–economic model, but for now I want to turn my attention to the "mind–body" problem.

As early as the seventeenth century, when we knew little about the brain and nothing about neurons, Leibniz stated that "mechanical reasons," based on Descartes's figures and movements, would never explain perception. This central thesis of the *Monadology* was repeated over the centuries, while figure and movement were replaced by chemical, neuronal, and synaptic interactions, introducing what was sometimes a dry brain, a kind of electric calculator, and sometimes a wet brain, primarily hormonal and emotional. This does not mean that we have learned nothing about the biological brain but that whatever neurophysiologists and philosophers have learned, it did not modify the terms of the problem they have always confronted. Some have arrived at a kind of methodological, even ontological, dualism, others have claimed that we are dealing with an unfathomable mystery, and still others have condemned, in the name of science, all the words used to describe the experience, fated to disappear the way phlogiston once disappeared.

From an ecological point of view, which is my point of view, the claims of dualism and unfathomable mystery are not much more interesting than the claim of condemnation. It's just that they account for what the latter abstracts: the fact that it is not easy to purely and simply disqualify, as simple opinion, the totality of words and knowledge about experience, motives, intentions. We are rather curiously attached to those words and that knowledge, and rather stubbornly inclined to agree with philosophers, psychologists, or other representatives of "mind" when they reply: "Make what we call motive, choice, intention as complicated as you like, but don't just try to eliminate them. Because then you also eliminate the possibility of saying why you—and you are the one who is arguing—harbor the intention to convince us that there is no intention." And yet, in the opposite camp, the full authority of the history of the sciences has been mobilized, as guide and guarantor. "As we discover the properties of circuits and systems and how they achieve macro effects, doubtless some time-honored assumptions about our own nature will be reconfigured. More generally, it is probable that our commonly accepted ideas about knowledge, reasoning, free will, the self, consciousness, and perception, have no more integrity than prescientific ideas about substance, fire, motion, life, space and time."[2]

That this confrontation has gone on unchanged over the centuries, ever since the coming into being of a physicalist point of reference, clearly shows that there is no "event" here or, correlatively, no practical invention. The confrontation fails to introduce any precise requirements or obligations. The person who denies the claim that "mind" can be reduced to what is most frequently referred to as a "state of the central nervous system" could just as easily be a Lacanian psychoanalyst, a phenomenologist in the Husserlian tradition, or a psychologist studying consciousness. The person who makes this claim has no idea of how this reduction might take place. In fact, nothing is more indeterminate than the identity of a "state of the central nervous system," and the behavior of such a state, which is to say, the "laws" it is supposed to obey and in terms of which thought would be explained. What variables are part of the definition of this state and what equations describe those variables? To what observables does it ascribe meaning, and what meaning? No one has the slightest idea how to answer these questions. The reductionists, who, in America, are curiously referred to as "materialists," limit themselves to putting forth faulty syllogisms such as: there is nothing "in" the brain other than physical–chemical processes; physical chemistry defines states and "explains" the behavior of a system based on those states; therefore, the concept of a "state" of the nervous system must constitute, for anyone who refuses to introduce a different type of causality, or invoke a "ghost in the machine," the far horizon of scientific research. In other words, even if the equations describing a state of the cerebral system can never be explicitly written, the "state" already has the power to provide a goal to scientific research, and already gives "materialist theoreticians" the right to question the validity of statements that seem to imply a "mind" irreducible to that state.

I want to characterize this situation as a case of "reciprocal capture." In Book I, I proposed reciprocal capture as a generic constructivist category that referred to a twofold construction of identity. Being generic, such a category doesn't provide a solution but introduces a problem: what is being constructed here? Certainly not a scientific practice in the experimental sense of the term—that is, neither the "brain" nor the practitioner, the one having succeeded in addressing the other in a way that invents both. What is captured is *any* result of neurophysiological research: every time the way some neurons interact is identified, every time a correlation is established between a neurotransmitter and some "modification of the state of consciousness," an additional step is said to have been taken that implies and confirms the goal of research. And what is fed and confirmed through this capture is an image of scientific progress, centered here on a fundamental claim, the claim to be able to deny, a claim directly inspired by physics. We don't know to what the brain will actually be reduced, but we already have a word to describe it, a kind of "code word." This is the word *state.* The denial of "mind" by the spokespersons of the unstoppable progress of science always refers to what appears as the crowning achievement of any objective science: characterizing its object in terms of a state, which defines at each instant the interplay of everything that contributes to its behavior.

Here, reciprocal capture operates as a matter of right: anyone who speaks in the name of a science, or the possibility of science, designates a reality she considers subject by right to the requirements of the approach for which she has become the spokesperson. Whereas the experimental laboratory creates what I have called, following Bruno Latour, "factishes,"[3] fabricated beings that are *as such* capable of an autonomous mode of existence, the one who claims the brain is reducible to a manifestation—although a very special manifestation—of the laws of physical chemistry seems to think that those laws are "there" in nature. In other words, their explanatory power would owe nothing to an experimental setting and everything to "reality." Correlatively, she herself is not subject to any obligation, she is merely the representative of a history that is itself authoritative. She is authorized to judge, and to condemn as part of a now disqualified fiction, any practice or knowledge that appears to create an obstacle to the manner of explanation that corresponds to that reality. Slowly, but surely, since only fictions slow its advance, the brain will be identified with a physical–chemical system.

How can we resist this reciprocal capture, which stabilizes a polemical image of science, on the one hand, and, on the other, is likely to hinder the construction of ways of thinking and of practices that can effectively evaluate what "studying the brain" might imply? That is the question I want to ask. To speak of "ideology," the imaginary, or rhetoric—in short, to trust the power of critical arguments—seems pointless. For centuries this type of argument has been used in various forms and always comes up against the same syllogism: physics has the power to explain anything produced in nature; the brain is part of nature. Therefore . . .

A first approach, already taken in *The Invention of Modern Science,* consists in considering the singularity of the power that is constructed in the laboratory.[4] More specifically, the difference between the power associated with experimental measurement and measurement in general, for instance, when instrumental observables are correlated with modalities of the "mind's" activity. Let us assume we accept the argument that nothing in the brain escapes measurement, and that every measurement is, or will be, so correlated. What does a measurement demonstrate, what does it signify, what does it authorize? It is the possibility, in some cases at least, of making such questions answerable in the laboratory that I identified with the invention of the theoretical-experimental sciences. Here, I'll limit myself to stating that experimental measurement implies that it is possible to recognize the measured phenomenon as a "reliable witness," capable of confirming the demonstration the measurement claims to authorize. Capable, therefore, of imposing obligations on everyone who takes an interest in it. The creation of a reliable witness implies an active mise-en-scène, or staging, which in turn creates a challenge for all operations of measurement: *not all measurements are equivalent.* And not all the data resulting from those measurements are equivalent either. We can measure anything we like, but a truly experimental datum must designate what is measured as being able to justify the meaning of the measurement, as well as define what must be included in the description and what can be eliminated as a contingent complication. Whenever no such staging has taken place, the description must be called empirical, or instrumental, in other words, determined by different types of instruments, no matter how sophisticated, and their corresponding observables. These observables commit us to nothing; the requirements they satisfy are those of the instrument, not those of a question that would supply its meaning to the instrumental datum. As such, they reflect the progress of instrumentation and are primarily governed by a dynamic of accumulation. We can assume that, one day, we will see happen to them what happened when we made the transition from observational astronomy to celestial mechanics, or from the empirical observation of epidemics to Pasteurian science. The work is not situated in relation to a goal in terms of which it would assume the meaning of gradual conquest, but in relation to the expectation of the creation of such a goal, through which we will finally understand what we have measured.[5]

It is also important to realize that, in any event, the concept of a "state" does not have any special association with experimental practice, in a way that experimental progress might presuppose and actualize the right to refer to it. For example, if we consider the staging of Pasteur's work, in which the microorganism and its transmission are the secret behind the spread of epidemics, or the breaking of the genetic code, which allows us to match DNA sequences with certain proteins, these two episodes, where experimental procedure demonstrated its power in biology, in no way allowed the construction of a "state," whether of the epidemic, the microorganism, or the bacterium. In both cases, the experimenter addressed a being, Pasteur's microorganism or bacterial DNA, and assumed it had the power to explain a phenomenon. In both cases, the crucial point is that the ability to explain means much more than a correlation among observables. The presumed agent has to preserve its presumed power under highly variable circumstances. It must allow the experimenter to "abstract" it, to separate it from its customary context of activity and present it under artificial conditions. In such cases, experimental staging requires a determinant relationship that is maintained through all the manipulations that test it, and it is this relationship's resistance to such testing that, reciprocally, provides it with explanatory power. But the relationship, precisely because it must be robust throughout a variety of experimental situations, reveals an "agent" not a state.

The neuron has now become an "agent," perfectly capable of demonstrating its functionality and role in neuronal transmission. But it is an agent within a population of neurons, with no determinate reference to the brain as such. The brain teems with agents, but the relationship between this multiplicity and the brain, in the sense that mental activity is related to it, is not defined. Neurons "explain themselves" through their relationships to one another but do not explain the "mind." It is under these circumstances that the state intervenes, as a presumed correlate of the "system" of all agents, an intervention that causes them to change their status from agent to component. In other words, the intervention of the state marks the transition from what the "agent" explains to what is to be staged, with the brain having the power to explain "mental production." Although connected agents are mute about their contribution to this production, reference to the "state" postulates that if all agents and their connections could be described at a given instant, they would provide the explanation of what we describe in terms of thought and feeling at that instant.

It should be noted that the concept of a state in the sense in which it could *effectively* produce a correlation between description and explanation has nothing to do with mere description. Let's take as an example the case of an expert who prepares an inventory of a house before it is rented. The inventory would be an accurate description of every relevant aspect of a house. This description has nothing to do with an explanation, however; it refers to the accounts the new occupant will one day be asked to provide. It constructs the possibility of a future evaluation concerning the difference between "before" and "after" occupancy, not that of understanding the process that will lead from before to after. The expert is limited to describing. And the limits of this description—which will be comprised in the "state"—refer to the possibility of damage, or the surreptitious substitution of one piece of furniture for another. The expert notes any existing defects (stains, cracks) and other features judged adequate for identifying the furniture. She does not record the size of the rooms, or their orientation, or how much sunlight they receive. This contrasts greatly with the questions that the "state of the central nervous system" generates if, by analogy with the states defined by physics, it claims the power to explain. Where do we stop when it's a question of describing the brain? How do we distinguish between what allows us to define it as "brain" and what would characterize it more contingently? Some theories of high-energy physics attribute a finite life to the proton and, in that case, a "cerebral" proton can disintegrate. Does this "count"? What about the fact that the chemical elements composing "cerebral molecules" can correspond to different isotopes? What are the variables, and what function defines how they are related? How do we describe, "at the same time," what our sophisticated instruments provide us with separately?

I don't believe that there has been any concept to this day that has been so misused, that has involved such disastrous blends of intuitive pseudo-evidence and an operation of disqualification, as the concept of a "state." And I suspect that the arguments I have raised will convince no one but those who are already convinced. For it is primarily intuition that gives its power to any argument that makes reference to this concept. *If we could* "see," describe, and characterize all bodies and their interactions, we would be able to discover the law that governs the whole much as we do in dynamics, deducing from the state at time t_0 any state at time t_i. It is with respect to this unassailable intuition that the critical arguments I have presented are powerless. In other words, they inevitably run up against the usual "yes, we know, but all the same . . ."

But maybe a change of interest, in the positive rather than critical sense of the term, could void the "if we could" of its fascinating power. This is my challenge. And my ambition in the pages that follow is to disconnect the "state" from the ideal of a finally objective, and therefore scientific, conception of reality, and connect it to a singular and rare construction, a significant achievement in the history of what we call physics.

I could repeat the same type of analysis for mathematical economics but I'll limit myself to indicating that in this case it's a question of abstraction rather than intuition that is at work. It is the power of dynamic formalism, its ability to "abstract" from the empirical specificity of situations when writing its equations, that fascinates economists. Here as well, we must try to introduce new interests, understand what the ability to write those fascinating equations requires. The challenge is to show to what extent this type of abstraction differs from the abstraction associated with logic, for example, or, more generally, from any language where a given word—such as *tree*—is capable of referring to a multitude: different plants, even artifacts that simulate the plants in question.

Logic takes the liberty of "deciding" to characterize a being in terms of abstract attributes. For example, by defining Socrates as a man, a man as mortal, and so on. It doesn't pay a price for this abstraction, wondering whether every human being will accept the relevance of membership in mankind, or if everyone will understand the term "mortal" in the same way. It need only conform to its own definitions, and not change them during its argument. Here, abstraction is a kind of unilateral generalization that ignores empirical specificity. I want to show that, when speaking of dynamics, abstraction means *singularization* above all else, an operation that exploits the singularity of what it deals with in constructing new forms of definition. Correlatively, abstraction in dynamics is subject to obligations: it must show that we are allowed to abstract, that is, that none of the questions that can be asked in the customary descriptions used in dynamics will be lost and that all of them will receive an optimal response, assigning a pivotal role to the singularity of the physical situation to which they correspond.

In other words, here, abstraction implies that the singularity of a situation, in the precise sense that it confers its relevance to dynamics, has been implemented and turned against the particularity of its empirical definition. It is this twofold operation of "implementing" and "turning against" whose invention must be tracked in order to show that there is no theoretical–experimental concept that is more demanding and less general than that of a dynamic state. The idea of extending to other fields the coincidence of description and explanation effected by dynamics could thus—hope is open to us all—join the world of hollow dreams to which it belongs.

There remains the question of the "faith" of the physicist herself, not only the way in which her example is captured but how she presents herself to us. The attempt to answer that question will have to wait. If it could be directly associated with this initial exploration, focused on the physical–mathematical invention of "dynamic systems" characterized by their states and by the equations that supply the law of their evolution, it would mean that the connection between invention and faith in physics has no history. When I conclude, I hope I will have shown that the vision of the world inspired by the notion of a dynamic system no longer enthrones physics in a position of authority, which we are required to accept or deny. It will have instead become a *question that stimulates interest* in the history that has framed physics in terms of such a vision. I want to change the nature of the question, which should no longer be about the world, about what the world requires of us through physics, but about the way physics has extended its own requirements over the world. Only then can I try to construct the historical problem that such a change of focus brings about: what happened to physicists?

8

The Singularity of Falling Bodies

"Seeing a body of a given size traveling with a given velocity, can we not estimate its force without knowing over how long a period of time and by what detours and delays it might have acquired the velocity it has? It seems to me that we can judge the present state without knowing the past. When there are two perfectly equal and similar bodies, having the same velocity, but acquired in one case by some shock and in the other by falling for a measurable duration, do we then say that their forces are different? That would be the same as if we were to say that a man is wealthier whose money took longer to acquire."[1]

The temptation to generalize the kind of judgment sanctioned by dynamics is not new. But no one knew better than Leibniz just how weak or circumstantial his argument was—the analogy between the state of a moving body that can be characterized by its "force" and the state of wealth, let's say, of a bank account. It wouldn't have been necessary to remind Leibniz—a philosopher of great subtlety, mathematician, and diplomat—of the difference between the perspectives available, even more so at that time, to the "nouveau riche" and to the holder of an ancient and honorable fortune. His analogy is not based on the authority of a definition of "wealth." It is intended to dramatize the relationship between what, for Leibniz, the term "force" implied and what the term "wealth" meant, assuming that wealth was defined abstractly, by the state of a bank account. The fact that this abstract definition was adopted by the majority of economic theories, which took physics as their model, would have probably caused Leibniz to wearily shake his head.

I have quoted Leibniz for two reasons. The first is that the above quotation situates the birth of the dynamic state within a polemical context. Leibniz published the above lines in February 1687 in Pierre Bayle's *Nouvelles de la république des lettres* in response to an attack by a disciple of Descartes, the Abbé Catelan. At the time, he was directly engaged in a battle with the Cartesians about the identity of the "force" conserved by motion: mv for the Cartesians, mv^2 for Leibniz. The second is that Leibniz is not only the "father" of dynamics in the sense that he was the one who baptized the science of motion initiated by Galileo, he was also *the* philosopher of dynamics, the one who best realized its radical novelty and extracted its conceptual dimension, presenting it as a "thought event."

It is sufficient to recall that Leibniz is the philosopher who harmonized apparently contradictory points of view. He was the one who stated the exact equivalence of a representation of the world "in which everything conspires," where everything that happens at a given point is determined by the totality of the world at that moment, and a representation where there is no longer a world but only a collection of noncommunicating "monads," the continuous change of each monad arising from its own internal principle. From a monadological point of view, the world we perceive and which we affect would then be a *fiction,* but this would be a "well-constructed fiction," for the perceptive experiences monads define, each for and by itself, are in harmony, as if they were different perceptions of the same world. Therefore, we can also say that each monad constitutes a distinct point of view about the same world.

Apparently, this is all nothing but philosophical speculation, which doesn't concern the physicist. And yet, as we'll see, Lagrangian mechanics will invent "fictional" forces and prove that the fiction is well constructed. And Hamiltonian dynamics will invent a representation of the dynamic system in which its various components, similar to monads, are described as following their own autonomous law, without interacting with one another. Was Leibniz a precursor? Did philosophy predate a history that specialists in rational mechanics have taken a century and a half to construct? It would be more fitting to say that Leibniz grasped, and speculatively extended, the singularity of the dynamic object, the same singularity that those specialists in rational mechanics would exploit technically.

The "force" Leibniz uses to confront the Cartesians is one of the concepts that helps harmonize, for Leibniz, distinct points of view. It is both physical and metaphysical. It is physical in the sense that in some situations it must be susceptible of measurement. It escapes physics in the sense that its measurement does not define the force's identity, which is based on the metaphysical definition (monadology) of the world created by god. Consequently, the question of measuring forces is not simply instrumental. Relevant measurement depends on a type of situation that *exhibits,* in a way that is intelligible and demanding, the rational harmony between physics and metaphysics. Measurement is always a fiction from the metaphysical point of view, but it will be a well-constructed fiction in situations where it exhibits an equality between the "full cause" and the "entire effect." The possibility of such privileged situations defines the very singularity of dynamics, and I'd now like to compare those situations with the one presented by Descartes.

In Descartes's world there were two kinds of physically intelligible situations: static situations, illustrated by the lever or balance, and collisions between bodies that are characterized as simultaneously hard and perfectly elastic. In the first case, intelligibility is limited to the device—balance or lever—at rest. If the balance is no longer at equilibrium, motion occurs. But Descartes doesn't believe it is possible to establish, in a way that is not "capricious," any relationship between the velocity of motion caused by disequilibrium and static weight, as measured by the equilibrium between two bodies, that is, by the immobility of the balance. The static measurement of weight excludes the consideration of velocity. On the other hand, whenever two moving bodies collide, their velocity is not caused but given. For Descartes, the collision exhibits conservation, and what is conserved is the product of mass times velocity.

However, after his death Descartes's followers had to confront an adversary who offered a unified view of statics and motion, where velocity was caused. What does Leibniz's balance reveal? Not the immobility of two mutually equilibrated bodies but the equality of their *conatus,* their respective tendencies, opposed by the device, to movement. Here, Leibniz follows Galileo, who had used a balance to identify the "propensity to movement" of a "heavy" body. The balance is no longer used to measure a "static" weight but what Leibniz calls "dead force," the tendency to motion that is prevented from occurring by the restraining action of the other body. Dead force can be used to identify the motion the body would experience if it were freed from its restraint, for it measures the infinitesimal acceleration that, accumulated over time, would result in increasing velocity, that is, an increase in "live force" (*vis viva*). Therefore, rest at equilibrium does not correspond to zero velocity but to an "embryonic," or "vanishing" velocity of indeterminable magnitude, to motion that is continuously being born and continuously canceled.

When Leibniz set about criticizing Descartes's "memorable error," he used the Cartesian balance as a specific case of the conservation of "force," something it wasn't for Descartes, because bodies do not have velocity at equilibrium. Leibniz, not

Descartes, was able to extend the conservation of the quantity of motion mv to the balance, the velocity then being "virtual," embryonic, infinitesimal. And he was then able to show that Descartes was "right" in this case because it was a very specific, purely accidental case, the only case where there was no reason to distinguish between acceleration and velocity. On the other hand, in the general case where velocity is actually found to occur, only mv^2 applies rather than mv.

It would be out of place to linger too long over the "querelle des forces vives" that, for decades, pitted Leibniz, and after him his allies, against the Cartesian physicists. The story is complicated for, although those in Leibniz's camp who supported "live force," mv^2, benefited from a position whose consistency is, for us, easy to decipher, having remained that of dynamics, the same is not true of the Cartesians. The Cartesians had come to accept and claim that the quantity of motion, mv, was also conserved during the uniformly accelerating motion of falling bodies. They were therefore willing to situate themselves in the problematic landscape set forth by Galileo and Leibniz, that is, to extend the conservation of mv to situations that Descartes, in whose name they were fighting, would have rejected. From our point of view, they were unquestionably "wrong" and the construction of a "symmetrical" narrative that enables us to recognize the legitimacy of their position demands the artistry of a historian.[2] What I will emphasize, rather, is the scandal of Leibniz's proposal, as Descartes's heirs saw it. What interests me here is the fact that it was for them an affront to the elementary obligations of rationality. This scandal helps us see, contrary to our habits, the singularity of the requirements the dynamic state satisfied, and what could thereby be claimed. It illustrated the strangeness of the measurement of the live force gained by a body because of its fall as proposed by Leibniz.

Although Leibniz, following Galileo, represented falling motion as accelerating over time, he associated the gain in live force produced by this fall with a measurement that defines it as *independent of the time taken by the fall:* this is the crux of the quote from the 1687 text, where he states that we can judge the present state "without knowing the past." I have already pointed out that Leibniz's economic analogy could not serve as "proof" for anyone. In fact, it becomes frankly humorous when employed in its more complete form. For the measurement of live force does not only imply that we can "judge the state of the present without knowing the past." It also states that this "present state" allows us to "judge" the future, also independently of time. Just as "force" does not depend on the detours and delays that characterize its acquisition, it can also be used to define that which it renders the body capable of, while being consumed, independently of the detours and delays that will affect such consumption. In economic terms, this characterizes an ideal barter situation, one without profit or loss. It is of little importance whether capital is used up all at once or placed in a bank, or in a woolen stocking, to borrow the Leibnizian metaphor, the "effects" of those different strategies should be considered equivalent.

It's easy to understand that the Cartesians would criticize this as an intellectual scandal, the arbitrary choice of a definition of force that allows a crucial aspect of a situation to be ignored, an aspect that should affect the concepts of cause and effect. To make the controversy somewhat more concrete, let's take the case where Leibniz challenged the Abbé François Catelan. Assume we have two bodies of the same mass, one of which has a velocity twice that of the other. Using the measurement proposed by Leibniz, this body's force is four times that of the other, which means that, in consuming its force, it will be able to rise to a height four times that of the other. But, objects Catelan, how much time will it take for Leibniz's fourfold force to produce this effect? Twice the time! For Catelan, the effect can only be said to be fourfold, and thereby imply a fourfold force, if it is produced in the same time as the simple effect. We don't have the right to neglect time when evaluating effect. And if we divide the "fourfold" effect by the twofold time of ascension, the result is only double, and therefore demonstrates that force should be measured by the quantity of motion, *mv*.

The argument is apparently plausible.[3] But it ignores the singularity of movement of falling bodies that Galileo had revealed and that Leibniz had radically exploited, incorporating it into the very principle of his definition of dynamic (live) "force." This singularity is none other than the power conferred by the falling body on the person describing it, the power to define the instantaneous state of that body. And it is this power that Leibniz fully exploited and that enabled him to ignore what the Cartesians defined as a rational obligation: we don't have the right . . .

Tell me where you are, tell me your velocity—this all seems very simple. But if the question is addressed to a falling body, that is, to an accelerated movement, it is being addressed to a body whose velocity is continuously changing. Its "instantaneous velocity" cannot be observed the way ordinary velocities can, it cannot be measured by dividing the distance traveled by the time taken to travel that distance. The body "has" a given velocity at a given moment but that velocity doesn't characterize any distance in any amount of time. How can we measure it? It wasn't until he was able to answer this question that Galileo became "our" Galileo.[4] And Leibniz later generalized his response through the use of live force.[5] The instantaneous velocity of a falling body is defined as the "effect" of its past, judged from a determinate point of view: tell me what height you have fallen from. And it is also the "cause" of a future, judged from an equally determinate point of view: I'll tell you how high you will be able to climb.

This is the core of the argument about live force, the terrible problem facing the Cartesians, who were dragged onto a battleground that was not Descartes's own. Velocity had changed its definition. The v appearing in mv^2 is the instantaneous velocity, whereas the velocity of Cartesian conservation is one that characterizes a uniform movement, which does not address the question of its cause. If the Cartesians had attempted to speak about "instantaneous velocity," they would have been defeated before they had even begun: they would have come up against Galileo's inclined plane, the first experimental device that gave its author the power of representing the way in which heavy bodies fall. It is the ball itself, rolling down planes set at various angles in different times, that demonstrated that, from the point of view of the thing the ball's descent made possible, all that mattered was its initial height.[6] It is the ball itself that obligates its interpreters to accept that the time it takes to descend is unimportant from this point of view, and that time is also unimportant when one is defining the effect: the ball is able to climb back to its initial height, regardless of the path it will follow and the time it will take.

In Leibniz's terms, we can say that the "active—or live—force" of the ball is evaluated by the height *h*, for it measures what the ball acquires in descending from that height, regardless of the path, that is, regardless of the time it takes. That same quantity of live force is gradually exhausted when it climbs back to that *same* height, against gravity. The live force gained during the descent makes it exactly capable of the return climb, regardless of the path or the time it takes to descend or ascend.

In modern notation, we write: $mv^2/2 = mgh$. This "formula" is more or less familiar to most of us, and we are accustomed, whenever physics is involved, to seeing an = sign joining a series of small letters. And yet the ability to write that = sign is in itself surprising if we consider that it does not articulate homogeneous quantities, like two weights that steady a balance, but seemingly disparate ones. Here, the = sign does not reflect the general power of equivalence, the power that gives equal value to two different commodities, for example. Rather, it reflects the singular possibility of equating what it defines as cause with what it defines as effect, and in doing so disconnects cause and effect from any reference to a particular past and a particular future of the moving object. Cause and effect are reciprocally self-determining. Cause is not responsible for effect, it does not bring about the effect: its identity is derived solely from the relationship of conservation it shares with that effect.

The dynamic equivalence between cause and effect defines what is cause and what is effect in an operational manner. Cause is what is exhausted in producing the effect, while effect, in exhausting itself, will in turn reproduce the cause. And this definition benefits from the most prestigious property capable of confirming its claims: not only does cause provide the true measure of effect, but the measurement is reversible. Effect, to the extent that it can in turn become cause and define as its effect the cause that produced it, shows that nothing has been allowed to escape description.

Whenever it is a question of measurement, in the ordinary sense, we have a considerable choice. I can count the hairs on a rat or shave it and weigh its fleece. The significance of this type of measurement, like those that involve the nervous system, is open to a wide array of interpretations and arguments. Consequently, it requires nothing of us. The privilege of the Galilean object is to allow us to escape this sense of the arbitrary, creating the possibility of a measurement independent of human choice. It was as if it had the power to determine how it *has to be* measured, to ensure that the definition of cause and effect, determined by their operational equivalence, is objective, independent of the person describing them. This is not a "cultural" or "metaphysical" idea, which designates what can be called a cause. The definition of cause and effect can claim its independence from any human choice, any preference, any subjective decision. It is objective in the strongest sense of the term, in the sense that it appears to be dictated by the object itself.

We can summarize, in technical terms, the singularity of situations that satisfy the requirements of dynamics: any change can be defined in two ways, as *temporal change* or as *change of state.* If we know (to stick with our Galilean example) the path of a falling ball—its movement down several inclined planes whose angles are known, followed by free fall, for example—we can determine the gravitational acceleration it will experience at every point of its fall. And we can then calculate the time this particular fall will require. The description of temporal change assumes the possibility of characterizing the acceleration at each point of the fall and does not explicitly require the equality of cause and effect. The velocity occurs over time and is not defined by whatever is exhausted in its production. On the other hand, the change of state is indifferent to the path actually taken by the ball and the succession of accelerations it experiences. It is determined only by the difference in height between the initial state and the final state, whether the transition between the two has taken a second or ten thousand years. Live force, mv^2, and what, ever since the early nineteenth century, has been called work, mgh, describe the movement from the point of view of the equivalence between cause and effect. In the nineteenth century these were called *state functions:* a change in one or the other was sufficient to characterize a "change of state," the transition between two states of the system regardless of how it occurred, based solely on the initial and final states.

The possibility of constructing a state function has nothing to do with a general approach corresponding to the concept of a state. It corresponds to the singularity of the dynamic state: the different states of a dynamic object are reciprocally determined in terms of "distances" that are not measured in the space in which movements occur but in a space that uses "cost" as a metric, the space in which a displacement is compensated by a gain or loss of velocity. In the Galilean case, it is the height differential traversed, not the length of the inclined plane, that counts. To be able to define a system by a state function is to be able to provide an unambiguous definition of what its "states" are: by its own functional description, the "state" explains the identity of the system to which it belongs.

Can we construct a state function? This is an example of an apparently technical question, yet one that is decisive from the point of view of the physicist's practice. For, as we will see, satisfying the requirements of such a construction determines the very possibility of making description and reason coincide.

In pure dynamics, whenever motion is defined as satisfying the requirements of an ideal that defines it as frictionless, the two descriptions, over time and through their state functions, may appear to be equivalent and the "state function" may appear to be a simple, alternative, "abstract" version of the spatiotemporal description. However, they are profoundly different. Only the description based on state functions assumes and exploits the singularity of ideal dynamic motion. We can, whenever we follow the change of a Galilean body over time, incorporate a damping term, representing the way in which friction slows the ball and reduces the increase in velocity. The equation is simply much more complicated and implies a series of empirical evaluations. For it is no longer a question of knowing only the gravitational acceleration at each point along the object's path but also the effects of friction, and the fact that its effects vary as a function of the velocity of the body. On the other hand, in the case of a description using state functions, the problem is not simply "more complicated." The difference between a situation with and without friction is dramatic. Friction, if taken into account, will eliminate any possibility of causal measurement. The body will never rise to its initial height. The instantaneous state can no longer be defined independently of its past, and the futures it can have will depend on the particular nature of the paths it follows. The "distance" between states loses its meaning. In short, the state function can no longer be determined. The singularity of dynamic states that the concept of a state function reflects requires conservative motion, without loss.

Is the state function an abstract concept? Yes, in the usual sense of the term, in that it satisfies "abstract" questions and not the kind of intuitive questions brought about by the sight of a body in motion. No, if we understand abstraction as a general point of view, neglecting the empirical multiplicity of the situations it comprises. In this sense, it is the concept of evolution over time that is abstract, for it applies to any body at all, whereas the concept of a state function can be used to grasp the singularity of Galilean motion, where the cause is conserved in the effect. The conclusion of the quarrel between Cartesians and Leibnizians is highly significant as it displays yet another sense of abstraction, here associated with the ability to impose its own condition upon a situation. In 1743 d'Alembert, in his *Traité de la dynamique,* commented that it was merely a war of words. What mattered was the way force was measured. Cartesian measurement, using the quantity of motion, was appropriate whenever force was measured by an obstacle that eliminated motion at a given instant, as when two bodies were in motionless equilibrium. Measurement using Leibniz's live force was appropriate whenever measurement was provided by the sum of the resistances that gradually counteracted motion, as is the case whenever a body climbs a slope, gradually losing speed. D'Alembert thereby concluded a controversy he considered distasteful for its reliance on metaphysical speculation. One had to be a metaphysician, like Descartes or Leibniz, to postulate a force that remains invariable in nature. D'Alembert's "rational mechanics" wanted nothing to do with metaphysics. Every term had to be associated with the possibility of measurement.

And yet, the simplicity of the solution is misleading, for it hides its own condition. If d'Alembert is right in judging the quarrel to be the result of a misunderstanding that reflects metaphysical blindness, it is because he sees only objects that satisfy the requirements of dynamics. If d'Alembert can provide a definition of force that could be called "positivist," one that relies on measurement alone, it is because the case that Descartes had considered quintessentially intelligible has disappeared from view: the collision in which two hard bodies *instantaneously* change velocity.

D'Alembert studied the question of collisions extensively and knew that the only collision that satisfies the requirements of dynamics is one in which two bodies with different velocities never actually make contact. Mimicking the collision, dynamics represents the two bodies as if connected by a spring that is continuously compressed during their approach and then relaxes when they separate with a perfectly elastic rebound. Any discontinuity disappears, and we return to the case of measurement by live force: the gradual slowing down of each body is measured by the resistance presented by the other body through the intermediary of the compressed spring. The collision, which for the Cartesian atomists was the quintessential method of transmitting motion, has lost its intuitive intelligibility. There are all kinds of collisions involving bodies that are more or less hard or soft, more or less elastic or inelastic. While the loss of motion occurring in each type of collision can certainly be measured, it is not the causal measurement that singularizes "rational mechanics." The collision has lost its status as an intelligible object, for the terms it unites are no longer defined by a relation of equivalence but by a mere empirical quantitative evaluation, similar to the one used to describe the temporal change that takes friction into account.

"If we could know the state of the nervous system at a given moment . . ." We can now see just what this innocent wish conceals. If a system is not defined by a state function, the ability to describe its state provides no intelligibility. It is a simple, mute snapshot from which nothing particular follows. Only the state to which a state function corresponds can provide the one who describes it with the power to align description, measurement, and explanation. This is a highly restrictive condition, even for the science of motion, for it excludes anything, such as nonelastic collisions or friction, that is a source of loss. Can we identify any interaction in the brain that might satisfy such a condition?

9

The Lagrangian Event

For the time being, let's put collisions and friction aside. We can also ignore chemical reactions and the entire field of what is known as physical chemistry. None of the phenomena we intuitively associate with the word *nature* satisfy the requirements of dynamics or *rational* mechanics. This raises a problem, however. The association between the "laws of motion," whether Galilean or Newtonian, and the strange notion of a "law of nature" appears devoid of any meaning other than "social," the arbitrary affirmation of a hierarchy with dynamics at the summit. If this were the case, the first challenge to which the question of an ecology of knowledge was addressed would amount to little more than a kind of criticism, a reminder of requirements that, in themselves, should serve as limits to the power of equivalence. But often we are too quick to adopt a critical attitude, which consigns to mere psychology, or ignorant arrogance, something whose novelty it is unable to recognize. Before we criticize, it would be worthwhile to escape the trap of generalities posturing as a critical mind denouncing human illusion.

"Nature is expected to adjust its comportment to the mathematics learnt in their school days by the members of some international committee for mutual congratulation and world-wide sight-seeing. They often boast that, like Antaeus, they keep their feet on the ground, stoutly refusing to fly off into the rarefied azure of 'pure' mathematics. Their activity, however, suggests rather the dance of an earthworm, who never lifts his head out of the mud."[1] The angry, contemptuous author of these lines, the physicist and historian Clifford Truesdell, is not a "critic" but a polemicist. He knows where to direct his attack: the mediocrity of his colleagues and the conformity of institutions. A mathematical physicist in the tradition of the Enlightenment—a century that also saw contributions from Bernoulli, Euler, and d'Alembert—he hammers out his contempt for the history that followed, a history in which, with rare exceptions, he sees the ethics of mathematical physics overcome by laziness, mediocrity, and routine.

For Truesdell, the problem of extending the mathematics that singularize dynamics to phenomena that do not satisfy its requirements simply reflects a betrayal of the obligations that supply value to the mathematization of phenomena. While the mathematical physics of the eighteenth century was the work of lucid individual creators, that which followed was for him the work of a science that was mobilized, animated by a unanimous faith in the authority of preexisting formalisms to which, willingly or unwillingly, phenomena had to bend. He accused it of an a priori submission to a prepackaged mathematization, the transformation of what that mathematics requires into judgment norms, the correlative absence of any obligation regarding the relevance of mathematical description, the ad hoc reliance on "physical" arguments to fill in missing details or correct inconsistencies of description.

"Once we recognize that a theory is a mathematical model, we recognize that only rigorous mathematical conclusions from a theory can be accepted in tests of the justice of that theory. . . . A proof mathematically strict except for certain gaps made plain as the sites for future bridges is not at all the same as a 'physical' or 'intuitive' argument which claims to be a proof but is no more than a drug to whirl us over high mountains and across deep gorges by illusion, illusion which drops us when we awake at just the point where we started."[2]

Mathematical physics can, therefore, act as a "drug" according to Truesdell. And he very precisely dates the *event* after which addiction becomes possible. The announcement of this event[3] appears in a well-known sentence by Joseph-Louis Lagrange. According to Lagrange, his analytic *Mechanics* transforms mechanics into a branch of "analysis," that is, infinitesimal calculus. This can be rephrased as: "Describe your system and I'll write the equations." The situation has changed. Mathematics is no longer associated with the invention of problems, it is associated with their "formalization," their formal representation as mathematical equations. In 1788, with Lagrangian mechanics, a new kind of physical–mathematical being came into the world, an authentic "equation-writing machine." When a physicist writes "let us address the Lagrangian of this system" (or later on, as we'll see, "let us address the Hamiltonian of this system"), she is still, apparently, dealing with description pure and simple. And yet, she has just introduced a being that possesses the paradoxical autonomy of a "factish":[4] a being endowed with autonomy because it imposes its own questions, guaranteeing their objectivity and conferring an entirely new kind of autonomy upon the physicist. This gives her the power to detach herself from the "intuitive," spatiotemporal description at hand and the ability to define the problem, where previously she had to confront a problem.

Here, I am suggesting that physical-mathematical factishes are a key to an "ecological" approach to the question I wish to pursue—that of a physics haunted by the "laws of nature" to which the "vocation of the physicist" defined by Max Planck corresponds. What I want to do in the exploration that follows is to present the ecological singularity of those factishes, the requirements they assume are satisfied and the obligations those requirements entail, in order to unravel what is customarily presented as a unified whole: the experimental fecundity of physics and the "discovery" of the fundamental laws nature is said to obey. I want to make the difference between the modes of existence of experimental and physical–mathematical factishes, that is, between the power of experimental invention and the power of physical–mathematical fiction, as interesting as possible.

Like Truesdell, I feel there was a "Lagrangian event," and therefore a "pre-" and "post-"Lagrange. As for the betrayal he condemns, although he is certainly entitled to make such a condemnation, I will not discuss it here because it implies a reference to transhistorical values—values associated with physical–mathematical creation—which may be noble or demanding but are unable to guide me in constructing an ecological approach to physics. It is the practical novelty resulting from the Lagrangian event and the values this novelty brought about that interest me. What Truesdell condemns, what he sees as the history of a decline, can be followed as the history of the creation of a new "psychosocial type" of physicist, a psychosocial type the event doesn't "cause" but helps make possible.

What does the difference between this "pre-" and "post-" Lagrange consist of in the driest, most technical terms? The empirical description of a dynamic system can combine a range of forces, Newtonian force as well as the force associated with a spring. And those forces can act on a system that is itself composed of bodies. Some of those bodies may be free, but others are not, being maintained, for example, at a fixed distance by a rigid connector. In all cases, a number of forces act on a system of bodies. But the effect of each force depends not only on constraints that are particular to the system (rigid connectors, springs, and so on) but also on the configuration of the system at every moment. In short, before Lagrange, the physicist had to consider a crowd of different cases that imposed a specific presentation on her each and every time she examined a system. After Lagrange, however, the varied multiplicity of these cases posed one problem and one alone: the construction of the system of equations that will represent them all in the same way. In other words, the effects, although distinct, of all the possible forces can now be defined a priori, and this definition gives the physicist the power to judge and to neutralize the empirical difference.

The Lagrangian event can also be defined as an "escape" from the Galilean laboratory, where bodies are subjected to what, after Newton, could be characterized as constant gravitational force, a good approximation in this case. Galileo's experimental invention and the power conferred upon the = sign that followed can then be seen as making it possible to measure the instantaneous effect of a force in this particular case alone. Let's return for a moment to Galileo's argument: If two interconnected bodies lying on two planes at different inclined angles are at equilibrium with one another, the ratio of the equilibrated weights is equal to the ratio of the acceleration that each of the bodies would experience, once the link connecting them is severed, when descending the length of its inclined plane. From Leibniz's point of view, the equilibrium between two interconnected bodies provides a measurement of their respective "propensity to motion," the *conatus*, or acceleration, they would have during the first moment of their fall once released.

This Galilean–Leibnizian measurement is in no sense a generalization. It is a "principle," an authentic "fixed point" by which, through the intermediary of static equilibrium, acceleration achieves the status of a measurable magnitude.[5] And there is nothing general about this principle. In fact, to use static equilibrium to measure gravitational acceleration, we have to allow that the "static forces" acting on two motionless bodies have the same value as the forces that, at every instant during the fall of each body, determine their respective acceleration. This principle is quite different from a logical postulate, for it introduces a requirement about the forces involved. The generalization of the Galilean measurement of acceleration using equilibrium will require that the effect of the forces be defined independently of whether the bodies on which they act are in a state of rest or in motion. This is what Lagrange referred to as "d'Alembert's principle." And, once armed with this principle, Lagrangian mechanics was able to escape Galileo's laboratory.

The requirement expressed by d'Alembert's principle is the same as the requirement that allows us to construct a state function. But the "force" of friction, whose effect depends on the relative velocity of the two bodies in contact, just as it prevents the construction of a state function, also restricts d'Alembert's principle. However, in this case the requirement assumes a much more powerful formulation for it does not exclude, but includes. And in particular it includes, that is, incorporates into *rational* mechanics, the scandalous Newtonian force of attraction between distant bodies: while frictional force, which is familiar and without mystery, does not satisfy d'Alembert's principle, the incomprehensible attraction at a distance does.

How does rational mechanics define a force? Lagrange provided a crowning conclusion to the work begun by Euler, the real author of what we refer to as Newton's second law, $f = ma$. It is important to remember that this type of equivalence could not have applied to the force of attraction as Newton conceived it. While its effects were measurable, it was not itself part of mechanics but rather reflected the will of god, the true author of a world penetrated by the contemporary action of divine power.[6] The Euler–Lagrange force, however, is defined by the acceleration that is its measurable effect. As for the relation that unites Newtonian force to mass and the square of the distance, it belongs to the world of empirical description. The force of attraction, together with other "forces," such as the restoring force of the pendulum and the forces of electrostatic interaction, now become "rational" because their definition, by satisfying "d'Alembert's principle," provides the force-acceleration equivalence with the ability to define them without having to take rest or motion into consideration. Consequently, rational mechanics moves from Galileo's laboratory to the limitless heavens, for d'Alembert's principle can be used to claim that we remain *in the same world,* the one in which the effect of forces provides the = sign with its power of definition.[7]

D'Alembert's principle makes a statement about values, therefore: from the point of view of dynamics, not all "forces" are equal. And the difference in value has nothing to do with rational judgment in general. It revolves around the question of the relationship between definition and measurement. The instantaneous effect of any force can be measured by the instantaneous acceleration it determines, $f = ma$; but only "conservative" forces that satisfy d'Alembert's principle provide the = sign associated with measurement with its power of complete definition. If we are dealing with a frictional force, the equality between force and acceleration remains valid. But this equality does not define the force because acceleration also depends on the velocity of the body on which that force acts at every moment. If, however, the forces are conservative, the equality allows us to characterize the instantaneous state of a dynamic system regardless of the velocity, that is, as if it were "frozen" in time. Of course, each body in the system has a velocity, but just as the effect, at a given moment, of the different forces applied at different points—that is, the different accelerations that will determine those forces—does not depend on that velocity, the instantaneous state can be completely defined by a collection of forces acting on a collection of points. What this means—and it is here that the power of the Lagrangian fiction comes into its own—is that the description of the instantaneous state can be construed as if it referred to a state of static equilibrium like that of the Galilean weight–counterweight situation. "As if" is the keyword of the Lagrangian event. We must now analyze the nature of the power produced by the fiction d'Alembert's principle introduces.

This power comes from the mathematical sleight of hand the = sign authorizes. Writing $f = ma$ is mathematically equivalent to writing $ma - f = 0$. The small difference between the two formulas is that the second *also* allows us to describe the equilibrium between a weight and a counterweight. The equilibrium is defined by the fact that the force, $-f$, exerted by the counterweight on the weight cancels the effect, ma, of the force associated with that weight. By analogy, every instantaneous state can now be defined by a fictional equilibrium at every point between the acceleration that determines the force at that point and that same force with its direction reversed. Some would say this is a rather convoluted way of describing a state, but the fiction has just found the fixed point from which it will be able to work in complete security. It is, indeed, this new definition of the dynamic state as a fictional state of equilibrium that will enable the mechanical problem to be reduced to a simple mathematical problem, that is, one that will bypass the work of invention necessitated by the variety of situations in which rational mechanics is relevant. Not by forgetting that variety, in the sense that forgetting would imply loss, but by translating it into a unique language that corresponds to the description of a unique kind of system, governed by unique kinds of equations.

If we consider each point of a dynamic system, together with the forces that are applied to it, we find diversity. The effect of each force at each point depends on the possible connections among the bodies that make up the system, the constraints on each arising from its being connected with other bodies or with all the other bodies. Only a system made up of free points, unconnected to one another, allows us to extrapolate directly, locally, from a force to its effect without having to ask specific questions about the constraints that have to be taken into account for the local definition of effect. But if those constraints could be incorporated into the definition of the forces that act on each point, the interconnected system could be described as if it were made up of free points. This new definition of force could be constructed because a state was assimilated to some fictional equilibrium condition between forces and accelerations.

In dynamics the = sign is a pivot point. If the instantaneous effect of a force is known, the force that determines it can be replaced by any force having an equal effect. The set of "effectively applied" forces can then be replaced by a set of fictional forces providing they have the same effect, determine the same accelerations at different points of the system. The fictional assimilation of a dynamic state to an equilibrium state allows a general definition of all the sets of equivalent forces, of forces that have the same effects as the actual forces on the system at a given moment. Such a set of forces is defined by the condition that if one reverses the direction of each force and applies them to the system at a given moment, they will act as a "counterweight" to the forces actually at work at that moment. In other words, they will cancel the accelerations the actual forces determine, creating a situation of instantaneous equilibrium. Just as (real) equilibrium had been the instrument for defining the relationship between force and acceleration for Galileo, Lagrange used (fictional) equilibrium to define, for a given system characterized by a spatial configuration of a given collection of point masses, all the equivalent sets of forces, that is, all the sets determining the same acceleration for each component of the system. And among those equivalent sets, Lagrange selected the one that allowed him to represent the system as if it were a system of free points, the acceleration and force at each point then being defined by the = sign. The Lagrangian transformation is defined by the fact that it allows the local equilibrium between (reversed) forces and acceleration to be expressed in terms of *independent spatial variables.* It amounts to representing a constrained system and the forces that determine its evolution as if it were a system of free points to which a set of fictional forces is applied.

The only subject of the transformation of representation constructed by Lagrange is the system as such. It is only with respect to the system and its diverse connecting constraints that equilibrium allows him to establish an equivalence of representation. In the context of Lagrangian formalism, to speak of a force applied to a point no longer has any stable meaning. The "local" definition in terms of fictional forces now implies all points and all forces. It is in fact a local point of view of the entire system. Similarly, the space in terms of which the distances between different constituents of the system can be determined has become *system space,* not a space we can measure in the ordinary sense. The transformation leading from the empirical system to the Lagrangian system represented as a system of free points, one that "interiorizes" the constraints restricting the freedom of movement of "empirical" bodies, effectively redefines spatial coordinates. Therefore, the Lagrangian transformation redefines everything that is empirically observable. Its only invariant is the fictional equilibrium that determines the instantaneous equivalence among sets of forces.

I won't discuss how Lagrange derived his equations here; the technique that interests me is the one used to formulate a problem, not solve it. And what is unique about the invention of the problem of dynamics by Lagrange is that it appears to make mechanics a "rational" language, devoid of assumptions, articulated in a fully transparent manner from unquestionable principles, free of argument, free of metaphysical convictions. This gave birth to the idea, so often advanced, that physics, the model of science, promulgates laws that ignore causality. The theoreticians of the social sciences, and economics in particular, who must constantly remind us that the correlations they establish cannot be compared to "causes," often use the example of rational mechanics to deny that, in doing so, they are giving up anything at all. As Bertrand Russell—who is endlessly quoted by economists—once wrote: "All philosophers, of every school, imagine that causality is one of the fundamental axioms or postulates of science, yet, oddly enough, in advanced sciences such as gravitational astronomy, the word 'cause' never occurs."[8] This is one of the consequences of the Lagrangian event: the ability to define mechanics as a neutral model, useful for all-purpose generalizations.

This ability is quite clearly deceptive. When the economist "represents a system" using "Lagrangian" equations, he obviously avoids having to determine "causes." He simply introduces "equilibria" and can, justifiably, claim that equilibrium is a neutral concept, independent of physical hypotheses. But he also exploits the definitional power of equivalence. If there is no longer any need for the word *cause* in Lagrangian language, if no specific statement corresponds to it, it is not because this language has gained its independence from the equality of cause and effect. By means of this fictional equilibrium, the definitional power of the = sign on which that equality depends has, in fact, been incorporated into the very syntax of the language of dynamics, that is, within the definition of the dynamic state. So we have equivalence, but it is "silent," the way a syntactic rule silently determines the statements it can be used to formulate.[9] We have not eliminated causality but succeeded in finally completing the transformation brought about by Galileo and celebrated by Leibniz: cause and effect are no longer categories of human judgment, satisfying our convictions about what can be called a cause or the choice of the point of view used to describe an object. The categories used to define a dynamic system are "objective," distributed by the = sign, which guarantees their legitimacy and ensures that they are measurable.

Until now I have limited myself to the instantaneous definition of state, where "cause" and "effect" mutually define one another through their instantaneous reciprocal cancellation. But what about the conservation of cause in effect, in the sense that it characterizes change over time? In other words, how do we make the transition from the definition of state to constructing equations of motion that define the trajectory of the system?

We have seen that description in dynamics correlates two ways of representing change: as a motion, which corresponds to the description of a trajectory defined in terms of space and time, and as a "change of state," which defines the "cost" of the transition from an initial to a final state, whatever the time and path associated with this transition. Lagrangian formalism generalizes this correlation. More accurately, it will make the first, which satisfies our space–time intuition, dependent on the second, which affirms the singularity of dynamics, the science of state functions. The idea that to "know" what a dynamic motion is all we have to do is watch a body fall or the moon move through the sky must now be abandoned. To "know" in the Lagrangian sense is to construct, and the Lagrangian construction does not define change in terms of movement in space or time. It defines it in terms of "distances" measured as costs, by what we now call *work*, the "price" of the transition from one state to another. Retroactively, Galilean height was already considered a measurement of the "work" performed against a Galilean force during a climb, regardless of the path. But work could not be used to describe motion, it could only express the submission of motion to the logic of equivalence. With Lagrange, the description of motion itself will be redefined as subject to this logic of equivalence.

To describe spatiotemporal motion so that the concept of change of state has the power to define it is to describe it as a continuous succession of infinitesimal changes of state, as a "displacement" leading from one state of equilibrium between forces and accelerations to another, infinitely close. This displacement results from the fact that, fictional or not, forces determine accelerations. And accelerations modify the velocities that characterize the different points of the system at that moment, thus contributing to a modification of its spatial configuration in the following moment. It is this new spatial configuration, and it alone, that, in the following moment, will determine the new value of the forces involved, that is, the new definition of the equilibrium among forces and accelerations to which this new state will be assimilated. The "motion" described by the Lagrangian equations is nothing other than the description of this succession of transformations of the definition of the equilibrium state that occur under conditions that are always different because, at every moment, the fictional equilibrium relates to a system whose spatial configuration has changed. So, every moment is "fixed," fictionally treated as the "first moment," indefinitely repeated, of a motion that is continuously reborn by breaking the previous equilibrium state, only to be immediately terminated through the re-creation of a new equilibrium state between redefined forces and new accelerations.

The motion described in terms of this continuous shift of equilibrium conditions no longer uses space and time for measurement but the cost of changing a spatial configuration, the infinitesimal work performed by fictional forces that correspond to that change.[10] The change in the spatiotemporal configuration of the system no longer "counts," except to the extent that it is "paid for."

Unlike the cost corresponding to the change in Galilean height, which was independent of the position of the body because Galilean force was constant, the cost of the Lagrangian change of state is obviously not given but must be redefined from moment to moment. How do we define this "cost"? A consideration of the forces involved could answer this question, but

Lagrange showed that another physical–mathematical construct can also be used. Of course, this construct is again a state function, for its variation characterizes the cost of the transition from one state to another. We call it the "potential" (or potential energy) of the system. Defining a system in terms of its potential is the culmination of what had been prepared by defining a state in terms of a fictional equilibrium. The system can be defined as the result of an aggregate of local interactions, but it is the system that determines the way in which those interactions must be defined. Similarly, potential can be defined as the result of forces that act locally, but the local definition of forces, from one moment to the next, can be calculated as the local derivative of global potential.

In this way, the Lagrangian system creates an "egalitarian" figure in the sense that, unlike weight for Galileo, no term in the description escapes being permanently redefined as a function of all the others. That force can be derived from potential means that what we call a force depends on the system as well as what we call a state. Temporal change itself becomes a description of the way the system continues to change its own definition of itself, characterizing its causes and its effects at every point.

D'Alembert, as we saw earlier, considered the argument over live force as no more than a war of words. Live force was suspect because it appeared to be a metaphysical law that transcended rational understanding, introducing a cosmological parasite into the purity of a language that ought to have relied on measurement alone. Along the same lines, Lagrange was very proud that "the conservation of live force" appeared in his *Mechanics* simply as a consequence, characterizing the solution of the equations but not appearing when formulating the problem. Leibniz's live force can, of course, be constructed from equations of motion: to the extent that the only forces allowed in the initial problem are those that satisfy d'Alembert's principle, that is, provide the = sign with its defining power, dynamic change can only be "conservative," can only conserve an invariant. In this case, like potential, live force is a state function of the system. Whereas potential is defined in terms of the system's spatial coordinates, live force depends only on the velocities that characterize the system at each moment. Conservation involves the sum of the potential values and the live force at each moment, and expresses, in terms of the state functions, the cost of the transition from state to state: what the potential function (V in Lagrangian notation) loses during such a transition is gained by live force (our kinetic energy, T in Lagrangian notation), and vice versa. And on the basis of T and V, Lagrange constructed a new state function, the Lagrangian $L = T - V$, which gives his equations of motion a remarkably concise form.

Is the Lagrangian definition "abstract" or "concrete"? And correlatively, was the intuitive idea of spatiotemporal motion that had to be abandoned concrete or abstract? I would like to make a brief detour here to differentiate physical–mathematical abstraction and logical abstraction.

Spatiotemporal motion can be imagined, and even described, by a series of measurements, regardless of the moving body: plane, planet, horse, bicycle rider. The moving body is "here" at one moment, "there" at another. This description, as intuitive as it seems, is, however, abstract in the sense that it is "detached" from any understanding of what led the body from here to there, and indifferent to what it is detached from. Starting from this description of motion, it is impossible to then characterize the being in motion—plane, planet, or bicycle rider. The description has been elicited unilaterally, and what it has overlooked cannot be regained. On the other hand, the definition of the Lagrangian system can be considered hyperconcrete, for not only does it overlook nothing, it fully defines the requirements that must be satisfied in the situations in which it is relevant.[11] It is true that the so-called abstract constructs Lagrangian definition introduces—potential, live force, the Lagrangian operator—are the result of a fiction that radically distances the terms in the description from the attributes we are intuitively tempted to associate with the different bodies that together make up the system. But the fiction enables us to create new attributes that produce a perfect coincidence between judgment and description. The Lagrangian constructs transform dynamical systems from a set of interacting bodies into a "being" with a priori well-defined properties.

Naturally, there is a cost to the power of this definition, and that cost is the radical differentiation among systems, depending on whether or not the interactions that characterize them satisfy "d'Alembert's principle." The slightest frictional force makes all the difference in the world, for it prevents us from defining the potential function.

What took place toward the end of the eighteenth century was indeed a change in power relations, marked by the coming into the world of the power associated with what might be characterized as a "physical–mathematical factish": a machine for establishing equivalence, endowed with the power to confer a unique form on the diversity of empirical situations. To this new being corresponds a new practice, endowed with the power to judge authorized by this unique form. And this new practice presents the problem of its value: the question of the pertinence of the requirements that a Lagrangian system must satisfy creates the problematic space in which the consequences of the "Lagrangian event" will be enacted.

In the eighteenth century, the *limits* of "rational mechanics," the singularity of the situations that satisfied the requirements of its "rationality," were explicit. After Lagrange, this singularity was able to fade into silence because it had been incorporated into the syntax of the associated equations. As we have seen, this syntax can even be used to claim that the connection between cause and effect had been made "obsolete" by physical rationality and to allow us to forget that the power of the = sign serves as the connective apparatus of the machine, the condition of possibility for "reducing" mechanical problems to a problem of mathematical analysis.

But the Lagrangian machine can only process what satisfies its requirements; it ignores the rest. And in doing so, it creates the question of defining the "rest" that escapes its grasp. The meaning given to this "rest" is critical. Either it attests to the singularity of the requirement that dynamics brings to bear on phenomena, or it is itself judged and disqualified as failing to satisfy those requirements. In other words, it is now possible to forget dynamic singularity to the benefit of some "norm": the only truly intelligible phenomena are those that satisfy the *conditions* of the equation. As for the "remainder," for whatever escapes the requirements of equivalence, it will then be associated not with obligations, but rather with something to be avoided—phenomena rather than descriptions will then no longer be of equal value. This is the situation Truesdell condemns. The specialists in mechanics have betrayed the physical–mathematical inventiveness that was the glory of the eighteenth century: "Nature is expected to adjust its comportment to the mathematics learnt in their school days" by those specialists.

10

Abstract Measurement: Putting Things to Work

I hope this initial approach has shown why I feel an interest in "technical details" is necessary. I have tried to free dynamics from the idea that it corresponds to a "vision of the world" in the sense of an intuition or belief. The power of dynamics is not one of vision, it is the power of the machine to create equivalences that subject empirical diversity to a unique problematization. That is why until now the term "determinism" has not been discussed. Of course, the Galilean object and the Lagrangian system are deterministic. This should be self-evident, and is part of the definitional power of equality. But there are many other determinist systems. The "real" pendulum, whose oscillation is gradually slowed by friction, is just as deterministic as the perfect pendulum. Even more so in the sense that, even if we ignore the details of its motion, we know that if we come back in a year, we will find that it has come to rest. Determinism is a general concept, central to "abstract" discussions where "if we could . . ." predominates, along with the rights it seems to allow. In contrast, the Galilean equality between cause and effect is a highly singular property, which has served as a fulcrum for an invention whose meaning does not refer to a vision or right, but to a mode of treatment.

What we have begun to follow is a history whose subject is the discovery of the transformations of physical–mathematical representations in which equivalence is the fulcrum, a history that as it unfolds invents new meanings for what we refer to as "abstract" and "concrete." Far from allowing formalism to win an extension, to give it the power of generalization ordinarily associated with abstraction, physical–mathematical inventiveness has distanced itself from all intuitive evidence by exploiting the singularity of its object, that is, by also creating a language that is exclusively suitable to it.

How can we relate this history? There is a Lagrangian before and after, of course, but the "Lagrangian" event does not determine its consequences. A new power relation has been established, but its consequences are not given. In another history Lagrangian formalism might have retained the status it had initially, that of a technical construct with a strictly limited scope.

No event has the power to act as a "condition" justifying the history of which it will become an ingredient. In fact, when attention begins to be paid to the event, each of the logical "links" in the great tale of progress begins to pose a problem. That is why the history that leads to the situation of reciprocal capture I described earlier, the history that sets up the scientist as judge and the state as that which generally allows us to associate "description" and "explanation," must be told. The key reference constituted by physics, the power, challenge, and authority of that reference in the field of science is certainly an ingredient, but it is to the precise historical configurations in which this ingredient appears that one must turn to discover the role it will play in each case. My objective is not to attempt such a narrative but to create an appetite for it. In fact, it is the only way I know to fight against the power of causes and the sufficiency of conditions: to awaken an appetite for different narratives that do not transform history into fatality.

We know one thing for certain: for the rational mechanic,

not all "effects" are equal; the only effects that are significant are those that are equivalent to their cause. Yet, among those situations that happen to be defined by their "defect," their "deviation" from this requirement, we find all those that are of direct interest to engineers. For the engineer, "dissipative" effects are not defects, for the requirements and obligations of her own practice are defined in terms of them. Just as the capitalist's problem is to maximize profits rather than respect the rules of a world of balanced exchange, the engineer's problem is not only to minimize the losses due to friction or shock, but, in certain circumstances, to integrate them positively into the design of a machine (a perfectly conservative mechanical clock makes no sense and there is no wheel in a frictionless world). The engineer also needs to organize the activity of workers so as to maximize the work they are capable of producing before they are physically exhausted, and, in this case, time counts: the worker's fatigue is different if he has to carry a five-pound load five times or a twenty-five-pound load all at once. All these questions, which are central to the engineer's practice, lack meaning in the world of the rational machine. From the point of view of this ideal world, all "real" machines are identified indifferently based on the "loss of work" for which their particular operation is responsible. The practical questions asked by the engineer are addressed to the empirical "remainder" that escapes the rational ideal.

The concept of work was formalized at a time when the social and professional identity of the engineer had just undergone a crucial mutation in France. In the early nineteenth century, the way engineers were trained was redefined by the government. The engineer no longer had anything in common with an artisan. He graduated from the École Polytechnique, where he was taught, first and foremost, rational mechanics as the basis of his practice. This implies that this practice was now identified as an applied science. The days are over when, as in the eighteenth century, engineers felt they were capable of entertaining a polemical relationship with rational mechanics. The term "mechanics" was cut off from its Greek roots, when it referred to the products of human genius, to marvels and clever contraptions. Ideally, work is conserved, but it is never created gratuitously. For us this may seem obvious. But when the Paris Academy announced, in 1753, that it would no longer examine proposals for a perpetual motion device, the clever inventors whose dreams were crushed responded indignantly. By the beginning of the nineteenth century, they had lost their public voice.

Every time a body moves in a space defined by a potential function, rational mechanics can measure "work," work being defined as the product of the displacement of the body by the force that resists it. Within the context of Lagrangian mechanics, work is a state function, just like the potential function V, live force T, and the Lagrangian operator L. Like the change in potential function, it supplies the "cost" of the transition from one state to another, and the work of the forces that correspond to that transition is independent of the path it takes. But this cost can be generalized outside mechanics. Provide me with a force, atmospheric pressure p, for example. Provide me with a piston that produces, against this pressure, a change V in the volume of an enclosure. I can tell you the amount of work performed: pV. Or, if you provide me with a worker who carries twenty bags of sand over a slope of ten meters, or twenty workers each carrying one bag over the same slope, I can tell you that the same amount of work has been done.

The concept of "work" became central to this process of generalizing mechanical reasoning. Work is referred to as a generalized mechanical measurement, "mechanical currency," as Navier wrote in 1819. But, as Coriolis noted in 1829, there is another definition of work: "it is what we cannot increase through the use of machines. . . . If we were to assume that we can construct frictionless machines, we could then say that work is a quantity that can't be lost."[1]

In other words, "mechanical currency," defined as something independent of its use, has little to do with the exhausting physical work of the laborer, or with its monetary evaluation. On the other hand, it has a great deal to do with setting aside the singularity of the dynamic object. Whether or not there is friction, or whether workers exhaust themselves carrying bundles, or whether steam moves the piston, we can always evaluate the mechanical work involved. But, here, work has lost the power of definition associated with a state function. It is no longer a physical–mathematical factish. It no longer guarantees that the transition between two given states has the same cost, regardless of the path. Work as a form of measurement in no way requires that workers exhaust themselves to the same degree, regardless of the pace of their work or the mechanical devices they use. It ignores this difference. It is applied to all cases where we can introduce the definition of a force, the one that provides its meaning to the device—piston or workers—that is going to be characterized in terms of its displacement against that force.

In this framework, then, work is a measurement *like any other,* determined by human interest, limited to providing a quantitative characterization of a situation that is not itself defined in those terms. Workers are "put to work" the same way as steam, and measurement by work defines their activity *unilaterally.* It is a measurement that demands nothing from what is measured and can relate indifferently to rats, men, rivers, steam, or a rolling marble, with or without friction. Here, we should speak, like Hegel, who was an expert in this type of distinction, of "abstract measurement" in the customary sense that abstraction refers to the selection of certain characteristics and overlooks others.

How are we to understand this stark generalization? Naturally, it has advantages because it enables us to estimate the work "lost" owing to mechanical imperfections, that is, to judge reality in terms of Galilean ideality. It also certainly corresponds to the general question of "putting things to work," which marks the nineteenth century, and the economic and industrial redefinition of activities as a whole. When a concept loses its singularity, it becomes a connecting point for multiple problems and practices and enables them to benefit from the prestige of its origins. Although work loses its constitutive relationship with physical–mathematical invention, the physical–mathematical ideality that constitutes the mechanical object that "works naturally and without loss" becomes the perfect metaphor for work as a natural activity, that is, "putting things to work" becomes a sign of progress. To eliminate any obstacles to "putting things to work" is to "rationalize," to purify this activity of the obscurantist matrix of custom, irrationality, and egotistical interest.[2]

Moreover, work, a measurement that is blind to what it measures, translates the condemnation of another possible. The difference between Galilean movement and real machinery could have become the site of an inventive practice, recognized as "scientific" just as rational mechanics was. A practice that might have succeeded in having those behaviors that are impossible without friction recognized as interesting in themselves and not only for their diverse specialized applications. Today, we can speculate about this history that never took place, for we are beginning to understand some of those behaviors. They have in common what is now called "nonlinearity" (the braking effect of friction, for example, depends on the speed of the body imparting the friction) and the consequences of such nonlinearity, which for a long time were no more than curiosities liable to interest only applied scientists, are now among the "major problems" of mathematical physics.[3] The question of the status of a particular form of knowledge, of its importance and interest, of its participation in the image a scientific field presents of itself, is connected with the hierarchies of the scientific community. The practice of engineering, as an applied science, was not entitled to present new problems that might have questioned the authority of rational mechanics. It has, however, created those problems, and has sometimes given them an elaborate mathematical form, but they have remained segregated in the field of engineering knowledge, incapable of intervening in the representation of bodies in motion transmitted by the great tradition of dynamics. Correlatively, the contemporary "novelty" constituted by the "discovery" of nonlinear processes reflects the fact that those involved with them today have, or demand, the social status of practitioners entitled to question the obligations and values that justified passing judgment on those processes as "merely anecdotal."

What I am struggling against is obviously the explanation of the aftereffect of the Lagrangian event in terms of some scientific "identity," whether historical (Heideggerian) or more concretely associated with the logic of capitalism, creating a generalized equivalence among commodities that determines their "exchange value" and allows us to describe exchange as circulation. Obviously, one of the characteristics of the event found in the definition of the dynamic object is the promotion of analogies between mechanical "cost" and economic cost. But if monetary equivalence had constituted the secret truth of the Lagrangian event, the repercussions of what would then be a "nonevent" would have been fully predictable: the equivalence would inexorably have been extended to all other fields, and physicists themselves, like hallucinating somnambulists, would have imposed the use of measurement by equivalence to everything they encountered. However, whether we are dealing with Leibniz's economic metaphor or work as mechanical currency—and there are many more examples—these references are not part of hallucinated belief (to the extent that they are not proposed by economists). Leibniz uses this reference in a polemic. French engineers are careful to point out the singularity of their "currency," its difference from other currencies used in socio-economic exchanges. And the sciences, such as mathematical economics, that have taken hold of Galilean equivalence have been at pains, from the beginning and still today, to carefully avoid interacting with the creators and true heirs of what they have co-opted for their own profit.[4] The analogy between dynamic conservation and commodity equivalence explains neither the Lagrangian event nor its consequences.

It is important, therefore, to avoid generalizations and relate the consequences of the Lagrangian event where it has effectively made history in the scientific sense, that is, *within physics.* It's easy to criticize mathematical economics, which "cranks" a Lagrangian type of machine with complete disregard for any obligation of relevance. But to disqualify the aftereffects of this event where the history of physics is concerned requires a commitment that affirms other possibles, like that of Truesdell, whose struggle distinguished the grandeur characteristic of physical–mathematical invention from the power of formalism. We need such a commitment because it is impossible to deny that the history through which dynamic singularity conquered the ability to stage physical phenomena, going so far as to plausibly claim they constituted an intelligibility "beyond appearances," is an inventive history rather than the blind repetition of arrogant pretense. That is why we are still concerned with the history of physics. We cannot stop to produce a critical verdict because this history is marked by the creation of problems that only the heirs of "rational mechanics" could formulate, problems that were nevertheless vectors of new practices, irreducible to equivalences that mimic Galilean measurement.

And here too, it is "technical details" that I want to address. The aim is to describe what has happened to *physicists* rather than just any subject of Western reason. And most of all, not to reduce this history to the monotonous repetition of a belief. The temptation is strong, however: Didn't Laplace already, with his demon, reduce the universe to the truth of celestial mechanics? And long before him, didn't Galileo say that "nature is written in mathematical characters"? In taking this step toward an ecology of practices, I am gambling that interest in the constructions that singularize physics is liable to make a difference, is capable of breaking the spell of continuity that makes Laplace and Galileo the spokesmen of the very identity of physics.

But a problem arises. Although I have used the "Lagrangian event" as my point of departure—the creation of a physical–mathematical device capable of formulating the problem of mechanical movement, that is, endowed with the ability to define what a mechanical problem is, what it requires, and what it justifies—narrating the consequences of that event is no simple matter. We could say that the question has been around for two centuries and is still relevant. And other ingredients, unforeseeable during Lagrange's lifetime, have arrived on the scene, primarily the question of thermodynamic energy, kinetics, and the quantum atom. What's more, the very way we are today able to inherit this history is open, subject not only to interpretation but to the divergent extensions of the physical-mathematical inventions that define it. How can we narrate a heterogeneous history without getting lost in obscure byways, and without imposing upon it a false sense of continuity, a simplification that would transform it into destiny?

The solution I've chosen, risky as it may be, is to take advantage of the "strata of contemporaneity" that specify physics. These are the strata that every visitor encounters in a university physics department. At one level, we come upon the direct heirs of Lagrange; at another, those who live and construct the consequences of that other defining event for physics, the "conservation of energy"; at another, history truly began when physicists were able to go "beyond phenomena" toward an unobservable elementary world. Naturally, instruments and references travel between these levels, but the requirements, obligations—the practices—are still distinct. It is this principle of stratification that I am going to try to explore. I want to devote the end of this first exploration to the direct heirs of Lagrange. No "discovery" that brings about a new practice and imposes the question of negotiation that this newness demands separates these heirs from Lagrange. That is why we will no longer encounter controversies here, for self-constructed objects are defined as subject to Lagrangian requirements. And yet, even here, Lagrange's heritage needs to be narrated twice, through two quite distinct though contemporary projects—that of Sadi Carnot (1824) and that of William Hamilton (1934).

11

Heat at Work

As we saw, the concept of work rather accurately reflects the institutional hierarchy that established itself, initially in France and gradually throughout the other European countries, between the proponents of rational mechanics and engineers. There are in fact two sides to this concept. The engineering side clearly indicates its subordinate status: the work effectively performed by a machine can be used to measure the losses characterizing that machine, that is, its deviation from the Galilean ideal. Work defined in this way can only be incorporated in evaluative and summarizing practices a posteriori, not as part of the invention of a problem. On the other hand, work understood as being ideally conserved is a state function. It is part of the arsenal of abstract concepts that allows mathematical invention to exploit the singularity of motion when the cause is conserved in the effect it produces.

And yet history has not respected the tranquil hierarchy I have just proposed. This hierarchy assumes that the real heirs of Lagrange will not be found among those who use work or any other mechanical function outside the field in which they are state functions defining their object. However, there is one example that escapes the opposition between unilateral "abstract measurement" and "intrinsic measurement," which remains true to the object's self-definition. And this is the physical-mathematical fiction invented by Sadi Carnot, who created the most astonishing thought object ever devised, a mechanical "pseudo-object" whose definition does not relate to any actual behavior but to the ideal of a form of control that would entirely determine its description.

We cannot replay history, describe the different paths it might have taken. But we can point out "improbable moments," when history, as we have inherited it, appears to result from a fragile combination of circumstances. Within the history of nineteenth-century physics, the role of Sadi Carnot, a young engineer who died prematurely and whose work was ignored during his lifetime, appears to be one such improbable moment. He was to become a key reference in the "crisis of values" that passed through physics during the latter half of the nineteenth century (see Book III, "Thermodynamics: The Crisis of Physical Reality"). But Carnot remains outside this crisis, his work belongs to the engineering tradition that devolved from rational mechanics. And—while not breaking with the past—he went on to provide this tradition with unexpected scope and significance, giving unanticipated meaning to the physical–mathematical object that satisfied Lagrangian requirements.

Carnot's invention involved a new type of machine, which first appeared in England the previous century—a heat engine. Following Napoleon's blockade of Britain (1806–13), French engineers discovered the extraordinary development industrial England had gone through and witnessed the explosive growth of output in successive generations of English heat engines. They had to make up for a significant delay, but they also wanted to establish the foundations for a rational description of the heat engine. Now, for the first time, French engineers addressed a problem that was not an "application" of rational mechanics but part of "cutting-edge" science, the science of gases and heat. A new natural force had been put to work, "fire," or more specifically the expansive power of the steam generated by fire. The English "monetary" evaluation (the English limited themselves to comparing the consumption of coal to the mechanical power produced, measured in terms of horsepower) was worthy neither of the event nor of French tradition. It was necessary to construct a theory for this work, to conceive an ideal machine that would allow for the a priori recognition and systematic implementation of the factors that would improve the output of such engines.[1]

Carnot was unique among contemporary French engineers not because of his choice of problem but in the way he formulated it. For Carnot, Galilean logic, which confirmed the conservation of a cause in its effect, would become an *inventive principle,* the creator of an object constructed to satisfy the requirements of the Galilean object. The device invented by Carnot is defined by the fact that it is, like the Galilean object, capable of dictating the way it must be described. But Carnot's invention is not experimental in nature. Unlike the Galilean ball rolling along an inclined plane, whose behavior can be used to identify cause and effect, Carnot's ideal device may be inspired by a heat engine but it in no way describes its operation, even purified. It is made up piecemeal and a priori to allow its author to claim that, in this case, the "cause" has been *forced* to produce an equivalent effect, exhausting itself in the process.

Therefore, in the case of the steam engine, the first question to be answered by Carnot was the identity of a "cause" that would, ideally, equal the effect—the mechanical power released. It certainly wasn't the burning coal, but neither was it the heat given off by combustion. Heat, for Carnot and his contemporaries, was a fluid, "caloric." Caloric is conserved and cannot be exhausted in producing an equivalent effect. What, then, "disappears" when a steam engine operates? Heat, or caloric, *at the temperature of the firebox.* The heated steam must be cooled so that the piston in the engine can descend and, therefore, the heat it contains changes from the temperature of the firebox to the temperature of the cooling water. For Carnot, the "cause" is the "reestablishment of caloric equilibrium." The caloric flowing from a "hot" source to a "cold" source "falls" from one temperature to the other, helping to restore the temperature equilibrium of the caloric. And the relationship between this restoration of equilibrium and the production of mechanical effect will have to be one of equivalence.

This is the major contrast between the objects of Galileo and Lagrange and those invented by Carnot. Once two bodies at different temperatures are in contact, caloric flows naturally from the "hot" body to the "cold" body, without any motive power being involved. The temperatures are equalized, equilibrium is reestablished, without the "cause" of the process, the temperature difference, producing any effect capable of restoring this difference through its exhaustion. Consequently, understanding the steam engine does not mean understanding the "natural" behavior of caloric but theorizing how a device might avoid this behavior and enable the reestablishment of caloric equilibrium to produce a mechanical effect. In other words, we can make an analogy between the potential difference defined by mechanics and the temperature difference of caloric, which, for Carnot, will be used to measure mechanical effect. Nonetheless, it remains that Carnot's "cause" is essentially different from Galileo's: any restoration of equilibrium "can" be the cause that produces motive power, but it is human artifice, the steam engine, that is responsible for actualizing that potential.

By what measurement and under what circumstances does Carnot's cause "cause"? The answer is simple: it is what Carnot's ideal cycle will introduce. "Since every re-establishment of equilibrium in the caloric may be the cause of the production of motive power, any re-establishment of equilibrium which shall be accomplished without production of this power should be considered as an actual loss. Now, very little reflection is needed to show that all change of temperature which is not due to a change of volume of the bodies can be only a useless re-establishment of equilibrium in the caloric. The necessary condition of the maximum is, then, *that in the bodies employed to realize the motive power of heat there should not occur any change of temperature which is not due to a change of volume.*"[2] In other words, maximum output excludes any change of temperature associated with the direct flow of caloric between two bodies at different temperatures and the challenge now became clear: Carnot had to invent some means to transmit caloric between a hot body and a cold body that avoided any contact between those two bodies.

How did the ideal Carnot cycle satisfy this drastic condition? Carnot used the then well-known experimental fact that the temperature of a body rises when compressed and decreases when it is allowed to expand. He limited himself to describing a cylinder equipped with a piston (ensuring a variable volume). The cylinder can be either thermally insulated or placed in contact with a "source" of heat. The "reestablishment of the equilibrium of the caloric" between the "hot" source and the "cold" source takes place in four stages:

1. The expandable body inside the cylinder—let's say it's steam—initially at the same temperature as the hot source, receives from that source a certain quantity Q of caloric without changing temperature. This implies that it expands and pushes the piston (*isothermal* expansion, or expansion at constant temperature).

2. Contact with the source is cut off but the steam continues to expand and push the piston. This implies a reduction in temperature. This *adiabatic* expansion (expansion with no exchange of heat) will continue until the temperature of the steam has reached that of the cold source.

3. This is followed by a new isothermal step during which the body is in contact with the "cold" source: the steam yields a quantity Q of caloric to the cold source at constant temperature, which implies that the steam is compressed.

4. During adiabatic compression, contact with the cold source is cut off and compression continues. The temperature of the steam rises until it reaches that of the hot source. The cycle is complete and can begin again.

Carnot's ideal cycle is purely and simply a fiction. We have no idea what kind of miracle makes it possible for heat to flow between two bodies at identical temperatures, or how steam can spontaneously be compressed while heating, or expand while cooling. In fact, no matter which step we examine, left to itself the engine would simply come to a halt. It functions only by deviating from the ideal. During isothermal expansion, an operator is needed to very gradually "heat" the system, that is, to maintain a temperature difference—we can make it as small as we like—between the hot source and the steam. Similarly, this operator must compress the steam—as gradually as we like—in order to "expel caloric" (step 3) and then reheat the steam (step 4). The last two steps, therefore, "consume" work, but this work is less than the work earned during the first two steps because isothermal compression takes place at a lower temperature than isothermal expansion.[3] For Carnot, the fact that the cycle is an ideal that can be approached but never reached is of no importance. In fact, he writes that this difference "may be supposed as slight as we please. We can regard it as insensible in theory, without thereby destroying the exactness of the arguments."[4] On the other hand, what is essential is that the ideal cycle be "reversible." It is able to function in reverse and transport heat from the cold body to the hot body. To do this it will use a quantity of motive power equivalent to that produced during its motor operation to transport that same quantity of heat from the hot source to the cold source.

In this way, Carnot showed that the "cause" that disappeared in producing the motive effect, the initial distribution of caloric between the cold source and the hot source, could be restored as the effect produced is exhausted. He created the = sign, which assigned cause and effect their respective identities, and in so doing demonstrated that his machine indeed provided optimal output. Otherwise, Carnot claimed, recycling the argument well known since Simon Stevin and used by Huyghens and especially Leibniz in his polemic against the Cartesians, we would be forced into the absurdity of accepting what is traditionally known as "perpetual motion." Perpetual motion, whose absurdity is postulated, is obviously not a motion that is eternally perpetuated, which is precisely the case with the ideal frictionless motions found in mechanics. It represents the perpetual—because free—production of motive power. This free production, Carnot demonstrated, would be possible if we could conceive of a machine whose output is greater than that of its reversible machine: all we would have to do is to operate it as a motor that generated motive power and use Carnot's engine to reestablish the initial distribution of caloric. To the extent that this hypothetical machine would be able to produce a motive power greater than that required by Carnot's ideal machine to restore transferred heat back to the heat source, a free remainder would exist after each cycle of operation of the two coupled machines. Absurd.

The Carnot cycle is a fiction but it is also a physical–mathematical invention of a new kind, a true "machine for creating equality" between two orders of phenomena for which no determinate relationship exists in the natural world. It is important to emphasize the word *create,* quite distinct from the term "purify." Mechanical idealization implies that whatever it addresses allows itself to be "purified," that the friction that dampens mechanical motion can "naturally" tend to zero (which occurs in celestial mechanics) so that the phenomenon, stripped of its parasites, can appear in all its luminous intelligibility. With respect to the Carnot cycle, the deviation from the ideal cannot be judged as a parasite whose elimination would purify ideal operation. For, as we have seen, it is the very condition of the cycle's operation, the reason why the system cannot remain eternally in each of the states it passes through.

The steam engine growls, vibrates, bangs, pants, whistles, hisses. Carnot's ideal machine operates in silence and smoothly, passing with perfect continuity from a state of equilibrium (during the distribution of caloric) to a neighboring state that is so near that we can describe the system as never having deviated from equilibrium by a finite amount. If this were not the case, it would mean that the condition made ideal by Carnot has not been satisfied: a change of temperature was determined by a direct flow of heat, not by a change of volume. Consequently, Carnot's ideal cycle *imitates* the trajectory of a Lagrangian system. As we have seen, the Lagrangian trajectory has little to do with the temporal deployment of continuously accelerated motion. That temporal evolution was subjected to the logic of equivalence, from which follows the very notion of a change of state (the same cost regardless of the path taken between two states). The Lagrangian trajectory is a passage from one fictive state of equilibrium, characterizing the system as such, toward another fictive state of equilibrium. However, there is a great difference between this trajectory and the Carnot cycle, because the movement from equilibrium state to equilibrium state described by the first is autonomous: the system itself supplies the reason for the change in equilibrium state that occurs at every instant. In the Carnot cycle, on the other hand, the ideally infinitesimal break in equilibrium is imposed by the operator, and it is therefore the operator who "imposes" a change by controlling what we can refer to as the "limit conditions of the system."

So, what did Carnot invent? The means to implement the logic of equivalence, to "rationalize" processes that basically escaped that logic: the flow of heat in this case but, later on, and based on the same principle, chemical processes as well. All the "thermodynamic" functions that characterize a chemical reaction, like the output of Carnot's heat engine, will depend on the construction of a fiction that mimes Lagrangian change. The actual chemical reaction will be replaced by an infinitely gradual "change in the state of equilibrium," driven and constrained by a change in limit conditions (pressure, temperature). And just as Carnot's ideal machine is never heated or cooled, in the sense that a flow of heat would result in a change in temperature, the ideal displacement of chemical equilibrium releases no heat.

Why is this type of mimesis important? To what does a science correspond that does not study processes as such but kinds of ghosts that have lost some of the essential qualities of those processes? It corresponds to a new definition of "reason," the kind that appears in the expression "rational mechanics." Carnot's device, like the "conservative" chemical reaction, is rational, but they satisfy a practice of *active rationalization:* the physicist no longer seeks to construct a purified description of the phenomenon she studies, she constructs from whole cloth a fiction susceptible to objective description, able to prescribe how it should be described. The measurement invented by Carnot is both intrinsic, because it points out the singularity of what is measured, and artificial, because what is measured is a fiction constructed to satisfy the requirements of rational equivalence.

We should not hastily conclude that a "state" in the broadest sense—generalizable to everything and anything—has finally found its source of legitimacy and that rational economics, for example, "does nothing different" than what Carnot's heirs do. Carnot's cycle belongs to physics, a science that explores the power of fiction, not the arbitrariness of fiction. The power of Carnot's fiction creates its own obligations. Whereas a "cause" in mechanics defines a state as a participant in change over time (through force or instantaneous acceleration), the cause presented in Carnot's cycle only has meaning because the system is driven by manipulation. As I have emphasized, the Lagrangian object had supplied work with the metaphor of a "natural" activity that progress limits itself to extracting from whatever paralyzes it. The object invented by Carnot refers to a kind of labor that we can require of any physical–chemical process providing we suppress its "natural" activity.

The active rationalization invented by Carnot, therefore, *obligates* us to introduce an operator, one who is actively responsible for the fictionalization, and it obligates us to recognize that the measurable fiction has no other temporality than that of its manipulation. And finally, it obligates us to remember that the manipulation in question is concerned with eliminating the type of behavior that "naturally" singularizes what is being studied (in the case of Carnot, the spontaneous passage of heat from a hotter body to a colder body without the production of "motive power"). The pseudo-Lagrangian shift from equilibrium state to equilibrium state serves as a thought experiment intended not to describe but to define the conditions that will allow behavior to be channeled, subdued, forced to submit to the equivalence of cause and effect. It does not introduce a relationship between cause and effect, it forces a cause to cause what it certainly *can* but generally does not cause. The equivalence between cause and effect no longer communicates with some notion of a "law of nature." It has become the attribute of a human device, invented to eliminate through manipulation whatever does not submit to such equivalence, to *make measurable* a phenomenon that does not itself determine its own measurement.

It is not irrelevant that at this point in the text, having established this contrast, I have been led to highlight the term "obligation," where I have, until now, spoken primarily of "requirement." The power of Carnot's fiction has a price, the difference between "discovering reasons," a victory that entails no obligation because the purification that made it possible will have eliminated only parasites, and "rationalizing," which obligates us to preserve the memory of the responsibility assumed by the scientist in establishing her object. In "Thermodynamics: The Crisis of Physical Reality" I discuss why this obligation is not insignificant and hasn't been viewed as such by physicists.

12

The Stars, like Blessed Gods

The reversible change of state invented by Carnot "mimics" a dynamic evolution, but mimics it the way a puppet mimics the spontaneous movement of a living thing. Conversely, with the equations of motion formulated by William Rowan Hamilton,[1] dynamics would come to share in the perfection of the ancient cosmos, of the circular movement of the stars conceived as free, spontaneous, and perfectly determined and intelligible. In 1876, the English physicist James Clerk Maxwell, whose poetic exuberance was unleashed after an exposé by his colleague Tait on the subject, rejoiced that dynamics was no longer the science of forces:

The shade of Leibniz mutters from below
Horrible jargon
The Phrases of last century in this
Linger to play tricks
Vis Viva and Vis Mortua and Vis
Acceleratrix:
Those long-nebbed words that our text books still
Cling by their titles
And from them creep, as entozoa will,

Into our vitals.
But see! Tait writes in lucid symbols clear
One small equation:
and Force becomes of Energy a mere
Space-variation.
Force, then, is force, but mark you! not a thing,
Only a Vector;
Thy barbed arrows now have lost their sting,
Impotent spectre!
Thy reign, O Force! is over . . .
The Universe is free from pole to pole,
Free from all forces.
Rejoice! ye stars—like blessed gods ye roll
On in your courses.[2]

We know that force is only a simple variation of (potential) energy, a derivative of potential with respect to space. But that the stars can be said to follow their course like gods, "free from all forces," is something new. And that the disappearance of forces to the benefit of potential can arouse such lyrical enthusiasm is also something new. No one had sung about the fictive forces of Lagrange or the virtues of his potential. Newtonian forces conserved their power in the heavens even when they were no more, in Lagrangian terms, than derivatives. Certainly, for Hamilton, Lagrange's *Mécanique analytique* had had the effect of a "mathematical poem," but it was the formalism he proposed that was able to impose itself as such on its users, not as practical fiction but as "true" in the sense that it confers order and beauty on dynamics.

This is not a minor point because the perfect harmony of the Hamiltonian equations, which proclaim both the multiplicity of points of view and their unity, will bring about a new type of value, creating a new kind of *territory.* To speak of new values is always to speak of new territory, and for those who inhabit that particular territory, the conflict of loyalty between belonging to a shared world where measurements take place in space-time, where bodies have weight and interact, and the beautiful truth of the Hamiltonian world, *becomes possible.* Shouldn't the physicist be free to select as "true," and no longer as a convenient fiction, those modes of description that express her problem in its simplest and most beautiful formulation? Conventional mechanical description might then be merely the darkened and deformed reflection of a primordial mathematical truth that provides movement with its luminous and intelligible simplicity. Forces would then be merely phenomenological, determined by our point of view of phenomena, and it is the syntax itself, the language that articulates the different points of view, that could alone lay claim to the value of an obligating truth.

A new type of realism here becomes possible, without which twentieth-century physics is incomprehensible. Lagrange's heirs diverge, for this realism now disqualifies, as a scaffolding that can be eliminated whenever the problem has been stated, all functions, force or work, that recall human questions and may serve as "common ground" between physicists and engineers. While Carnot's mime tied the identity of the system to the manipulation of an operator, Hamilton's mathematical poem would create a link between truth and beauty that in itself becomes a source of obligation, unbinding the description from contingencies such as observation and measurement.

What did Hamilton do? Lagrangian formalism retained a feature common to the changes that satisfied the requirements of rational mechanics and those changes in which "nonconservative" (or "dissipative") forces such as friction played a role. In both cases, temporal change may be defined in terms of independent variables describing the positions of the different points making up the system. Recall, moreover, that the Lagrangian fiction allows us to move from a description of a system of possibly interconnected bodies to a representation of that system that assigns it the same form as a system of free points. In technical terms, the Lagrangian representation of a system of N bodies has $3N$ degrees of freedom. Also, to the $3N$ position coordinates there correspond $3N$ derivative functions, velocities. These velocities are, therefore, *dependent* variables. Hamilton was able to show that this definition of velocity is unnecessary. Why not symmetrize the role of the variables of position and velocity, in other words, give the equations of change the responsibility for establishing the necessary relationships of interdetermination between position and velocity at each instant? Velocity would then become an "independent variable" along with position. Thus, the instantaneous state of a system of N bodies would be determined by $6N$ values of $6N$ variables without the value of any one variable limiting the value of any other. The system would then have $6N$ degrees of freedom. Where Lagrange derived $3N$ second-degree equations (acceleration implies the derivative of velocity, and thus a second derivative with respect to position) Hamilton derived $6N$ first-degree equations (acceleration is a first derivative of velocity).

But not only does Hamiltonian formalism greatly simplify the calculations involved (the first-degree equations are easier to solve than second-degree equations), it also entails a radical freedom of choice in the representation of the dynamic system with regard to the identity of the variables that describe it. The Galilean equality distributed what can be referred to as cause and what can be referred to as effect. This time, the equations distribute what are entitled to be called independent variables, which will then be referred to as "canonical variables" of the system. And it shouldn't be surprising to discover—playing the role of fulcrum for the canonical transformations that ensure the transition from a representation in terms of canonical variables to another representation—a new construct that both presupposes and embodies the power of the = sign. This construct is the *Hamiltonian* of the system. It is simply the energy (sum of Lagrange's potential energy *V* and the old active force, *T*, now known as kinetic energy) expressed in terms of canonical variables. A "canonical" transformation, generating a new representation of the system in terms of new "canonical" variables, must satisfy one constraint only: the Hamiltonian, expressed in terms of the new variables must retain the same value. As such, in the case of the Hamiltonian, canonical transformations achieve the most dramatic exhibition of the triumph of mathematical fiction over intuition.

Through the invariant character of the Hamiltonian's value, the = sign assumes a new power. It is no longer simply a syntactic principle of representation, it ensures the identity of the system throughout its representational changes. Each canonical representation that preserves the value of the Hamiltonian is, by definition, equivalent to the others, expressed in a different language. Mirroring Leibniz, the Hamiltonian articulates every possible distinct point of view for the same system. Like the Lagrangian, the Hamiltonian is also the key function in the Hamiltonian equations of motion. It is by means of the Hamiltonian that the changes of position and velocity are reciprocally defined. And—its crowning achievement—to the extent that any dynamic change by definition conserves the Hamiltonian (that is, energy), we can, if we wish, claim that change over time is itself nothing more than a continuous transformation of the canonical variables, a continuous change in an observer's point of view of a system that would itself remain invariant. As if it were the "observer" who was evolving over time, ceaselessly changing variables as her point of view changed. This, then, is what is implied by the seemingly innocuous technical statement "let us envisage a system as defined by the Hamiltonian." It announces that the physicist is in no way addressing a set of interacting objects, evolving in space and time, but a being that is so fictionalized that we can indifferently assign temporal change either to that being or to the way it is represented.

But the freedom of redefinition that detaches the identity of the dynamic system from the world in which humans prepare, control, and measure goes even further. The stars, Maxwell proclaimed, now go where they wish, like free gods. This would mean that the very role of forces of interaction, as represented by the potential function, has been reduced to an option, a matter of representation. It would be possible to define canonical variables in such a way that the Hamiltonian expressed in their terms, while of course remaining invariant, *consists only of kinetic energy.* This would imply that, described from this point of view, the dynamic system could be represented not only as a set of free points but as a set of points stripped of any mutual interaction, each following "its own path" as if the others didn't exist, like one of Leibniz's monads.

Such a representation does indeed exist. It is called a cyclical representation, wherein dynamic movements are represented in terms of periodic functions.[3] This cyclical representation, which will become the heart of quantum mechanics, may be considered the most complete expression of the power of fiction in dynamics. Causal equality was the point of departure for the adventure of dynamics. Here, it has been endowed with such power that any trace of a "cause" has been absorbed to the benefit of the triumph of invariance as such. Represented in terms of cyclical variables, the different degrees of freedom of a system evolve independently, each according to its own law, without the existence of one affecting the others in any way. The evolution of the system is nothing more than the evolution of a set of periodic "modes" independent of one another, the evolution of each mode being exclusively determined by its own initial values (and, of course, by the Hamiltonian from which the evolution of the variables over time is derived). But causality has been absorbed only by assuming the status of a definitional principle: the "potential" that represents the interactions between the components has been incorporated into the very definition of the variables. The change in each degree of freedom of the system is autonomous because its very definition already takes into account the totality of the system. This, then, is the perfect physical–mathematical realization of the world of Leibnizian monads, which can simultaneously be said to be *causa sui* and a faithful local expression of the universe they compose together.

Why is the representation in terms of cyclical variables so important? For two distinct reasons. On the one hand, it creates new beings of such great simplicity, such elegant autonomy that it is difficult not to be tempted by the idea that the transformation is *veridical,* that it gives expression to a "pure" reality, unshackled by the contingencies of our mode of understanding. The physicist then becomes a kind of Platonist: having left her cave and her distorting games with shadow and light, she contemplates a finally reunited beauty and truth. But cyclical representation also represents the triumph of the physicist in another sense: the possibility of constructing the cyclical representation of a system and the integrable nature of the equations that characterize it are, in one sense, synonymous. Once the Hamiltonian equations have been integrated, that is, once the explicit law yielding the variation over the course of time of each of the independent variables has been made explicit, the transition to cyclical representation becomes child's play. Conversely, if this representation can be constructed, the integration of the equations is child's play.

But can this Grail always be found? Can a cyclical representation always be constructed? If this ever turns out not to be the case, the road that leads from writing the equations to their solution cannot be determined. Will such an impossibility have the power to obligate physicists? Can't it be claimed, rather, that the problem is "merely technical," that the sun shines outside the cave, even if the road that allows us to escape from the kingdom of appearances remains impracticable?

I wrote that the cyclical representation of a dynamic system was the fullest expression of the power of fiction, but that is not its ultimate expression. The question of the integrability of dynamic systems suggests a new and extraordinary case of "physical–mathematical fiction." With the work of Henri Poincaré and the culmination of nineteenth-century astronomy, the dynamic system will, in effect, cloak a new identity, the first of the physical–mathematical identities that translate not only the defining *power* of equivalence but the problem of constructing the object whose behavior verifies that power. It is out of this new type of "problematic" identity that the question of obligations I have just raised assumes a precise meaning: what obligations will be associated with the obstacle to this construction identified by Poincaré? The question, which I will return to later, is still an open one for Hamilton's heirs.[4]

Is the solar system stable?[5] Are the disturbances the planets impose on each other's orbits, as determined by their interaction with the Sun (a two-body system), likely to accumulate, resulting in a catastrophic collision or the escape of a planet from its orbit? The question was first asked regarding the orbits of Jupiter and Saturn, which might cause some uneasiness. For Lagrange and Laplace, this question was even more crucial because the approximations introduced to calculate these disturbances had to result in calculations capable of facing a demanding challenge, namely, the confrontation with long-term astronomical data, and observations beginning as far back as the third century BCE! They succeeded, which allowed a much relieved Laplace to conclude that the solar system was stable. However, the problem was not completely resolved. For, weren't the approximations introduced by Laplace's calculations responsible for the stability attributed to the system? Today we know that they were and that the solar system is indeed unstable.

The fact that the question of dynamic instability was raised about the solar system is quite significant. Even if the time horizons it involves can be counted in billions of years, the stability of the movement of our planet is a problem that involves "us," one of those cases where the technical question of the effective construction of the trajectories of the bodies that are part of a system assumes an importance that cannot be silenced by the answer: "Outside the cave, where perfectly determined solutions satisfy deterministic equations, all dynamic systems are alike because they all satisfy the same Hamiltonian language." In a billion years, will the Earth continue to revolve along an orbit similar to the one it has now or will it have escaped into infinite space? This is the question that must be answered, not whether the Earth, like a blessed god, satisfies the equations of dynamics. The fact that the secret of its future trajectory resides in well-defined equations is not enough. The problem of its motion must be constructed in such a way that this secret can be revealed.

This is the challenge Poincaré's work will confront. He concludes that the expected answer was unlikely to be forthcoming and that the problem does not arise in celestial mechanics alone but constitutes the central problem of dynamics.

Poincaré generalized the approach of astronomers, who start with an "integrable" system, the two-body Sun–planet system, and study the perturbations caused by the existence of the other planets. And he generalized this approach not in terms of calculating trajectories but in terms of the very definition of the system, the construction of the equations. The "technique of perturbations" he employed took as its reference the cyclical representation of an integrable dynamic system. The Hamiltonian for this system, H_0, does not, therefore, contain any terms corresponding to potential energy and depends solely on the "action variables" J (a term used for the moment variable, mv, in a cyclical representation; see note 3). The definition of the dynamic system studied then corresponds to the cyclical definition of the integrable reference system *as perturbed* by the interactions that differentiate this dynamic system from the integrable reference system.

According to the new fiction constructed by Poincaré, the interactions characterizing a system are differentiated based on the problem of integration. Some of them, those incorporated in H_0, already characterize the integrable system of reference. The others, which correspond to the integration as problem, are represented by a potential function, V. This function, which describes the coupling between the degrees of freedom of the system, does not correspond to the identification of a state of things but to the formulation of a problem. This is why, regarding this new fiction, we can speak of the creation of a *hybrid* representation that associates what had until then been disassociated: the question expressed by the equations and the problem of the possibility of supplying an explicit answer to this question. Correlatively, this problem assumes a precise form: that of knowing whether or not the "perturbed system" can be defined in terms of new action variables, J', which could be used to define a new "free" Hamiltonian that depends only on these J' variables. If possible, a cyclical representation of the perturbed system[6] could be constructed and the problem would be solved; integration is once again no more than child's play and the cyclical representation of the new system will, if necessary, serve as the point of departure for a new use of the technique of perturbation for other systems. The ability to claim that dynamic systems are indeed integrable could thus be constructed bit by bit, because each "conquered" representation becomes the basis for a new conquest.

And it is along this path of gradual conquest that the obstacle of "Poincaré resonances" is encountered. Here, too, Poincaré generalizes the work of astronomers. At the beginning of the century, the astronomer Le Verrier[7] had already shown what happens when the relationships among some of the frequencies that characterize a periodic orbit calculated for the case of the Sun–planet system are "rational" (one is equal to or the multiple of another). In such a case there is a "resonance" between the frequencies and, as a result, even the smallest disturbance can have a disproportionate effect. However, in cyclical representation, the question of resonance, first associated with the periodic motion of planets, invades the entire field of dynamics. For, whenever we are dealing with a cyclical representation, every pair of degrees of freedom, represented as evolving independently of the others, exhibits periodic behavior, characterized by a frequency. From this point on the astronomical problem generated by the resonance between periodic behaviors no longer corresponds to a particular difficulty, which could be eliminated with a bit of ingenuity. It becomes a problem for the specialists of dynamics in general.

The questions of constructing a solution to the problem of integrating dynamic equations and that of identifying the obstacle to this construction thus present themselves at the same time. The very act of taking an integrable system as the starting point for this construction reveals the inherent threat: the frequencies that will present the problem of their resonances. In short, the general notion of "dynamic systems," which, since Lagrange and Hamilton, have all been characterized by the same type of equations, presenting the same problem, becomes clouded once it becomes a question of solving this problem, of moving from equations of change to the description of trajectories. To the extent that the definition of the dynamic object by a perturbed Hamiltonian is "hybrid," and already includes the problem of the construction of the behavior of that object, the equations themselves are qualitatively differentiated from one another, depending on the way they satisfy the requirements of integrability.

The discovery that the majority of dynamic systems cannot be integrated in the conventional sense of the term should not be seen as a failure but rather as a new success for the process of invention that we have followed throughout these pages and that I have tried to interpret from the point of view of abstraction and singularization. The representation of a dynamic system as "a perturbed integrable system," which can be used to formulate the problem of integration, cannot—and this point needs to be emphasized—be reduced to the procedure used by astronomers ever since Newton. In the hands of astronomers, this procedure corresponded to an intuitive construction of the problem: one began with a "Keplerian," two-body problem and calculated the perturbations associated with the presence of other bodies. But, with Poincaré, the representation of the "perturbed" system, that is, how the perturbation that couples the degrees of freedom of the integrable reference system is defined, no longer has any "intuitive" meaning. The perturbation has no meaning independent of the problem of integration. It is not an "interaction" that can be ignored as a first approximation but a deviation, which may or may not be absorbed, from the ideal represented by the construction of a cyclical representation. In no sense can it be interpreted as a "force" that would affect local behaviors. It cannot be conceived independently of the nonlocal, fictive definition that every cyclical representation embodies. In other words, the possibility of moving from a general recognition of powerlessness in the face of the integration of the majority of the equations of change in dynamics to the identification of what stands in the way of integration, once again illustrates the power of fiction, the power to reinvent the statement of a problem in terms that exhibit the singularity of what it is that allows the problem to be formulated.

13

If We Could . . .

What I have just described, with Carnot on one side, Hamilton and Poincaré on the other, are two different, even disparate, styles of reading the Lagrangian event, but not yet a history. For those heirs, for the moment, are not in contact. What type of problem could serve as the meeting ground for the active, "manipulative" physicist who follows in the tradition of Carnot, and the poet Platonist whose equations reflect the regular harmony of the heavens, who follows in the tradition of Hamilton? An ingredient is missing that would allow this history to begin, and that ingredient will force us to shift from the power of fiction to the controversies concerning the obligations that correspond to that power. But, to close this initial exploration, we need to evaluate the significance of the fact that there has been a history capable of bringing these two Lagrangian traditions into contact, a history that we know was concluded by the defeat of the "active rationalization" invented by Carnot, without which the notion that there are "fundamental laws of nature" would be met only with derision.

The confrontation—we can anticipate it since we now live in a world populated with atoms—cannot have been direct, the trajectory of the dynamic system versus Carnot's displacement between neighboring states of equilibrium. Every system fitting Carnot's model that we can manipulate and control is, in atomic terms, a "macroscopic" system. In terms of "actors" moving and interacting, it would correspond to a system characterized by roughly 10^{23} degrees of freedom. There is little point in trying to write down, even less of trying to integrate, the 10^{23} corresponding equations. A pointless dream, because the specialists of dynamics knew perfectly well, and long before Poincaré, that, without simplification, they were incapable of integrating the dynamic equations even for the general case of a *three*-body system. The confrontation, which resulted in the "physicist's vocation" defined by Planck, that is, which relegated Carnot's heirs to a subordinate status that identified them as "cave dwellers," had, in one way or another, to enforce a reference to something along the lines of "'if we could' integrate what we know we can't."

Every "if we could" reflects a decision about obligations. Its function is to state that between one simple reference situation—"when we can"—and another where this is not the case, the difference standing in the way creates no obligation, no inventive necessity, no new risk for the practice that requires the power expressed by the "we can." "It's the same thing, only more complicated." This expression always indicates that the relation of resemblance between two situations has been transformed into a principle of judgment. To that judgment corresponds the evocation of an imaginary practice that would actualize the resemblance and would thus cause the obstacle to disappear.

To refuse or deny an obligation is quite different from "believing," from being inhabited by a vision of the world. Refusal or denial situates us within an open history, as the rejected obligation could have been, can be, or will possibly be recognized. Believing refers to a much weightier causality that transforms history into destiny, and causes past, present, and future to resonate. From the mathematical characters that write Galileo's natural world to Laplace's demon and the triumph of Hamilton's heirs, it is the truth "of" the physicist, even—and why not?—the truth of a characteristic of human subjectivity in general, that both explains and is made explicit. The question of the future ceases to be "speculative," and limits itself to radical alternatives whose subjects become "mankind" or "reason."

This distinction is even more crucial since we must give credit to Hamilton's followers, and recognize the difference between their "if we could" and the "if we coulds" I have already questioned in discussing the "materialist" visions of the body or brain. The victory of Hamilton's followers over those of Carnot did not have (see Book III) as its battleground an "important" problem, such as the merits of a determinist or "materialist" metaphysics that would, for example, allow us to deny freedom or reduce thought to neuronal interactions. Naturally, the question of power is in play here, but that question isn't presented in "metaphysical" terms, distinct from practice. The confrontation between the two types of ideal powers invented by Lagrange's heirs—ideal power of control versus ideal power of definition—revolved around "cases" where a contrast between the two types of power could be constructed. In other words, the confrontation between the divergent lines of the Lagrangian heritage is part of the history of constructing the objects of physics.

Additionally, unlike the "if we could" of those who examined "the state of the central nervous system," the obstacle that the "if we could" of the specialists of dynamics has denied for so long as a source of obligation has been defined, as we just saw, with precision. This was done through the invention of a practice that succeeded in bringing into existence the meaning of "we can," thereby identifying the requirements that a dynamic system must satisfy in order for its equations to effectively confer the power to determine its behavior over time.

That is why Poincaré's definition of the criterion of integrability is so important. For the first time, the problem of requirement is not limited to confirming the conditions that would make the Galilean equality between cause and effect true (exclusion of frictional forces, etc.). It shows that, even if they all satisfy this primordial requirement, which defines the field of dynamics, all dynamic systems "are not equal." More specifically, they might not be equal if the obligations that correspond to the question of integration were confirmed.

To speak of obligation is to recall that an obligation is accepted but not demonstrated. If, over a long period of time, the answer to the question of how Poincaré's results obligate physicists has been "not at all," arguments were available justifying such an answer. It can be argued that, even if we do not know how to construct the cyclical representation of a system, nothing prevents us from claiming that it "exists." Therefore, it is only a question of recognizing a "technical" weakness the impossibility of defining a procedure that would guarantee the possibility of constructing the sought-for representation— something that does not challenge the identity of the dynamic system. Moreover, nonintegrability does not prevent us from relying on this identity for obtaining all sorts of results. For instance, statistical mechanics, which treats "sets" of systems characterized by the same equations of change, uses only the conservation of energy or the Hamiltonian invariance, and the very possibility of claiming that a set of systems characterized at a given time by the same energy value will conserve this characteristic throughout its evolution is the source of quite interesting properties with well-defined consequences. Finally, integrability in Poincaré's sense is the optimal property, but other mathematical techniques can be used to approach the behavior of dynamic systems. And, in particular, ever since the development of high-power computers, a method of inquiry has come into use whereby all dynamic systems are again equal:

the equation of change is not integrated but implemented by the computer. The machine "simulates" the system and produces, step by step, the trajectory the equations dictate. If the computer is not "obligated" by the question of integration, why would the physicist be? And over and above these specific arguments there is one that, ultimately, expresses the singular value of the Galilean factish: even if we cannot calculate it, the dynamic object *must* have a trajectory and one alone, the one that its equations determine, for these equations are not based on a simple description but translate and make explicit the reasons for the change. In short, because it is felt that the Galilean object and its heir, the dynamic system, are able to explain themselves, perfectly and completely, through their equations of change, the fact that the computer alone, in some cases, has the ability to understand this self-explanation, does not lessen its singular truth in the slightest.

Yet, there is a profound difference between the testimony of the computer, for which all dynamic changes are effectively of equal value, and the definition of dynamic behavior that the solution of dynamic equations offers. We can compare this difference to that separating an overflight of a tropical jungle from the path cut step-by-step with a machete by an explorer. The explorer may pass within three meters of a cliff, or a treasure, without realizing it. Similarly, a simulated trajectory may be qualitatively different from its close neighbors without anything in the simulation pointing out this detail. Every simulation discloses one and only one trajectory and is silent about its environment. Independently of the technical problems presented by any simulation, there is an intrinsic contrast here, for the definition of a particular dynamic trajectory obtained by integrating the equations follows from the general definition of any trajectory of the system. The integration of a trajectory cannot be dissociated from the definition of the landscape of all possible trajectories of the system. The simulated trajectory therefore dissimulates, beneath its appearance of exhaustivity, the loss of the power of definition of the landscape in which it is inscribed, the loss of the dynamic codefinition that integration brings about between a particular trajectory and the global landscape of possibles to which that trajectory belongs.

An obligation is accepted, not demonstrated, but we can define the implications of this choice a bit more precisely. There is an argument that states: "If we could define the dynamic trajectory whose equations hide the secret, we could verify that the 'nonintegrable' system does indeed behave in a way that confirms its membership in the same class as the simple systems that inspired dynamics." This argument transforms the experimental invention of the dynamic object into simple, contingent access to a truth that transcends it. For, if a dynamic language was created, it's because there were "privileged cases" whose behavior exhibited the singularity that conferred its operational power on the = sign. And, as we have seen, it is this operational power that was at work throughout the history of the construction of the dynamic equations. To behave as if the promise constituted by those equations, the promise of a solution that perfectly unites description and reason, was valid, whether or not it could effectively be maintained, is to abstract the experimental singularity of the effective objects of dynamics and unilaterally privilege what was gained because of that singularity.

In other words, trust in the "realistic" value of the physical–mathematical construction marks a contrast and discontinuation with respect to the fictionalizing process I have characterized throughout these pages, and in doing so transforms the meaning of power at work in the history of dynamics. The beauty of the power of fiction, as I have presented it, does not reside in the power of extension of dynamics but, on the contrary, in the invention of a way of formulating the dynamic problem that affirms its singularity in an ever more intrinsic manner. And the fact that this same power, the creator of formulations that are increasingly simple and elegant, also becomes a creator of difficulties is an inherent part of its demanding beauty. Independently of the power of fiction, the difference between "conservative" forces and "dissipative" forces (friction, viscosity, and so on) would not have had the radical significance it entails whenever dissipative force is defined as an obstacle to dynamic intelligibility. But the temptation to ignore the difference between integrable and nonintegrable systems gives precedence to intuitive evidence over the invention of the problem. Because these systems are characterized by the same types of equations, it appears obvious that the difference, which bears on the construction of the object defined by those equations, has no physical meaning. Such temptation amounts to justifying a separation between mathematical difficulty and physical interest, whereas the history I have related was marked by the creative power of mathematics as a vector of invention in dynamics.

Have I succeeded in taking a step toward an ecology of practices? Have I managed to weaken the intuitive power of the concept of a state, as expressed through different versions of the same refrain: "if we could fully describe an instantaneous situation (the neuronal brain, or even a society), we could deduce its behavior over time"? Have I brought into existence a new interest in the question "what happened to the physicists?" Have I turned into a question the fact that so many physicists, from Max Planck in the past to Steven Weinberg today, are able to claim that nature is indeed subject to laws characterized by the perfect and harmonious symmetry of the Hamiltonian equations? In any event, I hope to have created the means to establish a difference between the reductionist foolishness of the neurophysiologist or the "materialist" philosopher and the "faith" that animates physicists—a crucial difference for an ecology of practices, as with all differences that touch upon the question of power. The physicist who speaks of the "reality" of physical laws is not a reductionist. She is not inhabited by the passion to disqualify or defined by the reciprocal capture between the power of the so-called "objective" scientific process and the reduction of what exists to a mechanistic or physicalist "causality." On the other hand, we could present her as being *captivated* by the beauty of what the invention of dynamics has brought into existence.

If one of the implications of the ecology of practices is to bring about the mode of presence among us of different practices and their practitioners, this distinction is important. Without it, it is hard to understand the inventive character of the practice of those physicists who, in the name of a unitary vision of a world intelligible in terms of laws, forge increasingly audacious categories with great freedom, increasingly at odds with shared concepts of space, time, and causality. We must interpret Hamilton's experiment, his discovery of the "mathematical poem" constituted by Lagrange's *Mechanics,* in the strongest sense possible. The faith in mechanics and, more specifically, in the body of laws that have inherited its mathematical structure is not a simple "ideological" drift of experimental invention. It does not reflect the temptations of the power brought about by the reduction of nature to a vulgar automaton. It is addressed to a creation whose primordial characteristic is beauty. To denounce Planck's faith or Weinberg's would be inappropriate. To reveal their methodological errors would be redundant. Rather, it is as *poets* that we should address them.

Nonetheless, they are strange poets indeed, for the power they have of asking questions that, by right, should be of interest to all humans, of making discoveries on our behalf, and announcing the truth of the shared world, obviously constitutes one component of their passion. For that reason, we cannot simply call them poets and must follow the way in which such power has been constructed. But the best antidote against the fascination exercised by the power of physics may be to give them a presence as creatures captivated by their own creations, by the aesthetic harmony of physical–mathematical factishes whose power they experience, and to point out the difference between the power of fiction brought about by the = sign and the power of the reduction that denies and disqualifies.

BOOK III

Thermodynamics

THE CRISIS OF PHYSICAL REALITY

14

The Threefold Power of the Queen of Heaven

In 1873, the English physicist James Clerk Maxwell wrote to his colleague Peter Tait: "But it is a rare sport to see those learned Germans contending for the priority in the discovery that the second law of thermodynamics is the *Hamiltonsche Princip*. . . . The *Hamiltonsche Princip* the while soars along in a region unvexed by statistical considerations while the German Icari flap their waxen wings . . . amid those cloudy forms which the ignorance and finitude of human science have invested with the incommunicable attributes of the invisible Queen of Heaven."[1]

The German physicists, here Rudolf Clausius, Hermann von Helmholtz, and Ludwig Boltzmann, fought for the honor of who would be first to demonstrate that the law of increasing thermodynamic entropy was derived from Hamiltonian mechanics. And Maxwell laughed, or sneered. He knew that the characteristics of dynamics, the Queen of Heaven, the science of celestial trajectories, were *incommunicable.* Hamiltonian dynamics is indifferent to the statistical considerations through which the Germans attempted to associate it with the cloudy forms of thermodynamics. A phenomenon defined in terms of thermodynamics cannot satisfy the requirements of mechanics.

Maxwell's sneer doesn't surprise us. For Maxwell knew what

we know, or think we know, and what the Germans apparently did not. But what is important to understand today is how some believed, for years in fact, that they could confer upon a law of thermodynamics the status of a statement in dynamics.

And yet, that sneer of Maxwell's also reflects a new situation, an authentic ecological mutation. In Book II, "The Invention of Mechanics," I followed the diverging paths of the two traditions that followed from what I called the Lagrangian event: the construction of equations whose syntax affirmed the power of equality between "cause" and "effect," a power that became the creator of a fiction, the constructor of an "object" that defined, according to its own terms, space, force, and movement. Hamilton marks one of those paths and Carnot the other, each of them extending, by mutually incompatible means, the power of the Lagrangian fiction. Yet, the two protagonists introduced by Maxwell, the Queen of Heaven and the second law of thermodynamics, seem to have overcome this incompatibility, and this holds true not only for the "German Icaruses" but for Maxwell as well.

At the time when he wrote to Tait, Maxwell had already invented a small but "very observant and neat-fingered being" capable of sorting the molecules that we—ignorant beings with finite capacities—can only describe "in mass." In 1874, "Maxwell's demon" was born once and for all in the words of his colleague William Thomson. And the function of this demon was to show that "the dissipation of energy," whose inevitable nature the second law of thermodynamics proclaims, is in fact the result of our finitude and ignorance. "It is only to a being in the intermediary stage, who can lay hold of some forms of energy while others elude his grasp, that energy appears to be passing inevitably from the available to the dissipated state," wrote Maxwell in 1872.[2]

Maxwell's demon is still with us: his presence is explicit whenever a physicist addresses a nonprofessional audience, but it is implicitly required whenever it is necessary to introduce an approximation that allows us to proceed from a "fundamental level of description" to an "observable property."[3] And on every occasion, Maxwell's Queen of Heaven seems to proclaim her triumph, the triumph of the physics of laws over the physics of phenomena. We have indeed entered a new history, one that confirms what I have previously called (see Book I, "The Science Wars") the "psychosocial type" of physicist, who is identified by a *vocation:* the ability to transcend disparate phenomena and their associated operational know-how and access a unique and objective vision of the world, independent of human prejudice and interest.

The first expression of this vocation is found in Maxwell's letter to Tait. For the Queen of Heaven "soars along," she is indifferent to statistical considerations (the procedures of approximation that confer meaning on experimental properties). She enables her spokesperson to disqualify, for having created a false resemblance, the way in which the followers of Carnot subjected the transformation of heat and movement to the syntax of the Lagrangian equations.[4] The Queen of Heaven is therefore a vector of obligation: we must not confuse a science marred by ignorance and finitude with the purity of her reign. And the value judgment she authorizes effects a radical hierarchization of the two hereditary paths of the Lagrangian event.

But, one might object, is this situation really so new? Doesn't Laplace's old demon bring about the same hierarchization? Isn't Maxwell's demon the same thing in a new context? Don't both demons entrust probability with the responsibility of articulating our uncertain world and the "objective" reality ruled by law? Yet, from the ecological point of view, the *identity* of a being includes the way it forms relationships with other beings.[5] From this point of view, the similarity between the two demons is greatly attenuated.

Laplace's demon, which appeared in the introduction to his

Essai sur les probabilités of 1814, had as its principal, and even sole, function to ensure a peaceful coexistence between the deterministic world of the laws of motion to which it had access and all the situations where probabilities impose their relevance upon us. Laplace's demon did indeed affirm a hierarchy, but it was a hierarchy from which no particular consequence followed for physicists, and which imposed no obligations upon the users of probabilities.[6] He annoyed the philosophers, but caused no controversy. He assumed a place in a rather old history characterized by the philosophical conflict between atomists and Aristotelians concerning the nature of movement, or the theological question of the possibly privileged relationship between matter as devoid of final causation and the intelligibility of the world as a divine creation.

Maxwell's demon, on the other hand, introduces a hierarchy *within* physics, a discrimination between different practices of knowledge that are addressed to the *same phenomena* but correspond to distinct requirements and obligations, between which a value judgment can operate. The creation of a hierarchy of this type constitutes a crucial moment from the ecological point of view, and the event takes place at a precise moment. It would not have been possible during the first half of the nineteenth century and, one way or another, its consequences would be stabilized only during the early years of the twentieth century, when the great theme of the physicist's vocation was being introduced. In the case of Maxwell, the Queen of Heaven, already invoked by Laplace's demon, confirmed her power, and that power was threefold.

The power of the Queen of Heaven is already manifest in the fact that the "German physicists" criticized by Maxwell appeared to have succumbed to her seductions. While Maxwell was able to satisfy himself with a technical finding—the "attributes of the Queen of Heaven" are incommunicable—the Germans were led to believe it was possible to "really" extend those attributes to the world of thermodynamic phenomena. She triumphed a second time through the value judgment that claimed that thermodynamics is based on an artificially created resemblance with dynamics: the Queen of Heaven has the power to disqualify what, for some physicists of the period such as Pierre Duhem, represented a conquest of rationality, the abandonment of any "metaphysical" claim in favor of a sober and lucid relationship between rational description and experimentation. And, with Maxwell's demon, she finally claimed the power of a vision of the world. The "incommunicable attributes" of the Queen of Heaven would then have been communicated to the population of molecules the demon manipulates so that all forms of energy are equally available. The jurisdiction of dynamics appears to have been extended to all the phenomena described by thermodynamics.

The threefold power that singularizes the Queen of Heaven is new. Contrary to Laplace's demon, it has nothing to do with an all-purpose rhetoric but reflects a vocation with consequences for the physical world and the physicist. Here, we confront a crucial moment from the point of view of an ecology of practices. Whenever a modern scientific practice is limited to judging or disqualifying what has gone before or its environment, as Laplace did with his demon, the endeavor reflects the preexistence of hierarchical relations but does not present other problems. The only interesting question concerns the future, when those relations may be called into question, when the ignorance and arrogance they made possible are no longer confused with the power of reason or science to disqualify appearances (and especially opinion). As for the past, it is simply a question of not taking it seriously any longer and learning to laugh at those who honor it. However, the Queen of Heaven presents a very different problem, which binds past and future in a very different way. The power she claims was indeed embodied in a program for physics, a program that, for better or worse, generated new practices indissociable from twentieth-century physics.

How can we "state" the existence of a neutrino? That is the "symbolic question" presented in Book I, "The Science Wars." For the neutrino has a dual mode of existence, experimental and theoretical. Its experimental mode of existence does not, in fact, create any special problem in terms of an ecology of practices, any more than the microorganism that Pasteur brought into existence, or any other "experimental factish." It "simply" affirms the singular autonomy of the beings our experimental practices bring into existence, the singular requirements that whatever we fabricate in the laboratory must satisfy before it can be said to "exist." On the other hand, unlike Pasteur's microorganism, the neutrino is characterized by a theoretical mode of existence that cannot be disassociated from a factish of a very different kind, one that is much more fearsome, a true engine of war that judges and hierarchizes. It is this power that we see at work in the disqualification of thermodynamics. And it is the fabrication of this very particular factish that needs to be followed step-by-step. For the principal challenge, here, is to avoid directing the narrative toward its eventual historical outcome, toward the interpretation introduced by Maxwell's demon. It seems that the Queen of Heaven has been endowed with the power of defining the truth of all natural phenomena, of escaping the singularity of dynamic systems her "incommunicable attributes" make explicit. But where does this power come from?

We need to frame this problem based on obligations that correspond to an ecology of practices, that is, without endorsing the result, but also without condemning it in the name of consensual norms. Also, it is important to avoid depriving it of its interest, for example, by referring to all-purpose "macrocauses" such as the irresistible attraction of a determinist understanding or the imperative seduction of explanations that extend "beyond phenomena." For the "ecological" possibles whose existence is at stake have meaning only if they meet the challenge of acting as vectors of new interests. Here the ecological possible entails the production of different modes of relation among the various contemporary protagonists of the triumph of the Queen of Heaven, whether these are physicists who proclaim their faith in her, other practitioners who are judged by her, or the "public at large," for whose fascination she has been presented. I want to try to create interest, through the *question* of her triumph, in all the different possibles that the history of this triumph has silenced. To the extent that those possibles are not foreign to the problems posed today by physics, they might prepare the ground not for the conversion of the physicist, but for a moment of hesitation. To make room for hesitation is not to suggest new paths for contemporary physics. New paths are not mine to envisage. They belong to those who would create them.

15

Anamnesis

All narration, if we are not careful, will follow the slope that leads back to ourselves. It ignores uncertainty because it knows the answer. In this particular instance, because it knows the new identity assumed by twentieth-century physics. I, on the other hand, want to "slow down" the movement, make interesting the moment when the various ingredients of an ecology of practices come into play: requirement, value, and obligation. And I especially want to "slow down" the transition from the problem of values that created a crisis in physics at the end of the nineteenth century, to the apparently "purely physical" solution that would later be supplied.

I want to begin by recalling the questions with which Kant was able to assign limits to the field of positive understanding. The definitions of what humankind can know, of what it must do, and of what it can hope for were to remain outside this field for all time. Yet, the problem of value presented to the physicist at the end of the nineteenth century could be formulated similarly. What should physicists do when they discover that all physical-mathematical representations are not equal? Can they hope to wipe out the ignorance and finiteness revealed by thermodynamic representation? What becomes of rational knowledge if they invoke a knowledge that is not their own, but that of the manipulative demon? Must they accept the obligations that herald their ignorant finitude? The questions Kant hoped to use to block the invasive ambitions of positive knowledge are taken up again and repeated, although differently, within physics—what can the physicist know, what should she do, what can she hope for? We must slow down the history that would inspire the arrogant notion that such questions might find a solution within physics.

We need to first avoid giving this return to Kantian questions a grandiose interpretation, one that might confirm, for example, that physics does indeed touch the boundaries of knowledge because it discovers therein the need to bend before the inescapable interrogations of philosophy. Consequently, we can also state that this return itself reflects the fact that there is nothing specifically philosophical about the interrogation in question, that it can be formulated within any practice. In other words, the Kantian questions would not be addressed to "man," naked and universal, but would involve the particular obligations associated with a practice. All in all, "naked and universal man" is found nowhere except in the practice of the philosopher who has actively purified humankind of any attributes judged to be secondary, mere obstacles to the requirements of universality that, or so it would appear, identify philosophy, and are therefore unworthy of imposing obligations upon it. Transhistorical questions such as "What can I know?" "What should I do?" "What can I hope for?" "Who am I?" are part of this definition of philosophy. They are all that remains after one has purified everything. I would go so far as to claim that their apparently inescapable character derives from the fact that they are all-purpose versions of practical questions, that is, questions related to practices, to their territory, and to the movements of deterritorialization that affect them.

To return to the moment of uncertainty when the practical

identity of physics was at stake is not an archaeological undertaking as Michel Foucault understood it. It is not a question of bringing to light forgotten strata or of resisting the temptation to judge them in terms of the "not yet" that defines history as progress. Nor is it a return to a Freudian or genealogical past, which could be used to reexperience the present as the rejection, repression, or recovery of something that would continue to be repeated everywhere in disguise. Rather, the approach is one of *anamnesis,* here defined as the voluntary evocation of the past. For, if the past has been forgotten, there has been no disguise but rather the intent—then the habit—to forget. This is shown by the fact that the texts to which we should return remain perfectly readable, and we could even say that their readability is disturbing. In their case, the principle of symmetry between "winners" and "losers" taught by the contemporary history of science does not require any form of heroic asceticism. The questions asked by these texts have not been suppressed in such a way that they would be repeated involuntarily, nor have they been canceled by a conquering knowledge that would drain them of meaning. It's as if there had been a "decision" to turn our backs on them, a decision to withdraw from the obligations they make explicit.

As a voluntary evocation, anamnesis reflects an approach that is far from neutral. It is a question of establishing, in the present, obligations whose deliberate rejection becomes—based on this reading—part of the identity of twentieth-century physics. Needless to say, this approach is imprinted with a certain violence. It does not participate in the respectful sharing of the stated values and emotions experienced by those it describes. But the physicist's "faith" in the intelligibility of the world proclaimed by Planck, or the enigma of that intelligibility—which so amazed Einstein—will not be denounced as base dissimulations or sordid manipulations. The deliberate nature of the rejection does not indicate that physicists were aware of the meaning I would give to that rejection. Planck, Einstein, and their successors did not deliberately seek to escape the problem of their obligations. Based on my reading, they rejected a perspective they felt would endanger the physics to which they were attached, the physics that made them think and hope. In other words, their rejection cannot be separated from a historical situation that associated the question of obligation with conflicting positions.

The anamnesic approach is not a denunciation or deconstruction. Naturally, it implies acknowledging the serious impact of the notion constructed by Planck, then Einstein, that couples the physicist's faith to the enigma of the fecundity of that faith. For both faith and enigma situate the physicist's obligations as if they referred to an elsewhere, to a world whose intelligibility *must be made to exist* outside phenomena, and to which no one may have access if they don't already share the form of engagement that defines the physicist. Ernst Mach, against whom the weapon thus forged was used for the first time, was quite aware of the polemical nature of this definition of physics: "After exhorting the reader, with Christian charity, to respect his opponent, Planck brands me, in the well-known biblical words, as a 'false prophet.' It appears that physicists are on the way to founding a church; they are already using a church's traditional weapons. To this I answer simply: '. . . I renounce with thanks the communion of the faithful. I prefer freedom of thought.'"[1] But the anamnesic approach does not imply that we follow Mach in the contrast he creates, in which he makes himself out to be, opposite Planck, the heroic defender of freedom of thought. We need to be cognizant of the fact that the confrontation among physicists went so far as to explicitly include the question of the relationship between reason and faith. But it is not a question of taking advantage of this extension of the polemic to take sides or to bring the question onto the more familiar terrain of so-called ideological quarrels. The very extension of the polemic is part of the event whose memory is to be constructed, and this extension cannot be used to explain or interpret it.

Anamnesis is not a critical deconstruction because it does not belong to the register of solution or explanation. It belongs to the very continuation of an event whose heirs include everyone who believes that the "neutrino exists" (including me). How does this existence obligate us? Is it possible to discuss, to negotiate something presented as a whole, the experimental neutrino and the neutrino whose claims are bound to those of the major laws of physics? The anamnesic approach must recreate the problem through the byways of the history that has produced its solution, a history of winners as well as losers. But where does its ability to do so come from?

In the health-care professions, anamnesis reflects a certain trust on the part of the therapist in her own ability: she should be able to read, in varied and apparently disparate manifestations, the unity of a disturbance whose diagnosis will guide her subsequent therapeutic approach. We immediately note the ambiguous nature of this proposal. What is this trust based on? On the intrinsic power given to the therapist by knowledge she claims and which enables her to identify the "one truth" hidden by multiple appearances? Or on her skill in being able to produce, in the performative sense, a unity capable of guiding her? Ordinarily, the second interpretation is part of critical, or relativist, commentary, challenging the proofs invoked by the first and revealing the circularity between the conviction that guides the action and the action that confirms the conviction. But there is nothing necessary about the polemical nature of this second interpretation, and the practitioner who accepts it would not thereby become a disheartened "postmodernist," reducing every claim to truth to a disenchanted "it works." This would be the case if and only if there had been agreement concerning another point that, until now, has remained implicit. For the performative production of a unity is, quite obviously, not the business of the therapist alone: the person with whom the operation of anamnesis is conducted cannot remain indifferent to it. The question then becomes one of knowing if we can identify an operation that confirms its own presumptions with the closing of a circle. This implies that the person with whom the practitioner works is indefinitely available for any operation of this type, the way a sandy beach on which we follow the trace of our own footsteps is available for any imprint. The situation would be different if whatever an operation addresses were capable of making a difference, that is, if it were not equally available for all operations. This then introduces new questions, especially the question of the obligations entailed by the therapist's practice.

We will return to these questions later. The current problem relates to the approach I intend to follow in the field of the history of physics. To the extent that the different positions currently at odds with one another are not at all in need of therapy and that, even if this were the case, those who hold such positions would certainly not acknowledge that I am likely to fulfill the role of therapist, the risk implied by the metaphor of the sandy beach is not very relevant. Physicists are anything but ready to confirm simply any interpretation of their science. The analogy would rather be that of a soldier traveling through a field laid with mines, some of which have already no doubt exploded. Readers who have followed me thus far should be aware of the fact that others have already closed this book with a shrug, rejecting some of my propositions the way Planck and Einstein rejected those of Ernst Mach. And yet, here as well, anamnesis assumes a certain trust, trust in the ability to recognize as noncontradictory positions that have succeeded one another in noncontradictory ways, to recognize them as contrasting facets of the same problem.

Obviously, that trust cannot be dissociated from my own standpoint regarding the problem presented by the "incommunicable attributes of the Queen of Heaven." That trust I learned, in large part, from working with Ilya Prigogine, who reinvestigated what had been considered by many to be a closed book in the history of physics. Although I'll provide a description of Prigogine's work later,[2] it is already present here in my discussion of the Queen of Heaven, more specifically in my discussion of the requirements associated with her attributes.

Does this mean I am attributing to Prigogine's statement of the problem the ability to discern the "one truth" dissimulated by multiple appearances? And in this case does it mean that, for me, all of the problems that created a crisis in physics in the final decades of the nineteenth century lead to a single problem that Prigogine solved, as if by a miracle? A "solved" problem, a crisis overcome, closed, and, as such, comprehensible? This is not at all what I am proposing. On the contrary, I maintain that even if Prigogine's solution came to be generally accepted, it would not constitute a solution to the "crisis" as such but a *rebound,* the creation of a proposition that both accepts the historical outcome that followed this crisis and reactivates its scope and consequences. In other words, my reading is consistent with a proposition that extends the history of the crisis in physics because it is my interest in this proposition that led to my interest in this crisis. But this reading in no way implies the confirmation of that proposition. What interests me is the fact that it was able to be formulated. Moreover, the *ecological* meaning of such a possible rebound remains an open question for me. Even the (very) hypothetical acceptance of Prigogine's proposition would not resolve the problem I'm after. My goal is not to look for the conditions of consistency between dynamics and thermodynamics, between the physics of laws and the physics of phenomena, but to explore the requirements and obligations entailed by an ecology of practices. Among scientific practices, physics holds a place that is certainly unique, but we must avoid conferring a privileged status on its questions and crises.

In fact, the situation is more complex. The "critical" questions, to return to my paraphrase of Kant—what can the physicist know? what should she do? what can she hope for?—were certainly present in the work of Prigogine and his collaborators. And it was by following the sometimes labyrinthine pathways of this work that I learned to understand its practical meaning and scope, creators of values and obligations for the physicist. But whenever a scientific problem culminates in a solution judged to be satisfactory, whenever it has finally defined and identified the landscape where its requirements can be satisfied, it also creates, in keeping with that landscape, the simplest, most unequivocal, most adequate way of formulating itself. The labyrinth is replaced by the straight line, questions are replaced by the possibility of stating what Maxwell and his contemporaries "did not yet know," and which explains the difficulty they had to confront. In this way the time of science—"peculiarly serial, ramified"—is constructed, "in which the before (the previous) always designates bifurcations and ruptures to come, and the after designates retroactive reconnections."[3]

The moment the function and the state of things to which it refers are mutually actualized is a moment of intensity and risk, when the life of the scientist is suspended, when night is always too long, interspersed with doubts and torments. Does the reference hold up? Will it resist the challenges that match its claims? Do matters of fact adequately respond to all the requirements of the function? Do these requirements have absurd consequences? Yet this moment is also one directed at a future in which it will be pointless to preserve the memory of the multiple components of the problem that has finally been resolved: "the function of the scientist's proper name is to spare us from doing this and to persuade us that there is no reason to go down the same path again: we do not work through a named equation, we use it."[4] In other words, this is the moment when a perspective comes into existence in which what is mixed together can be separated. A new, immaculate state of things separates out from its history, becoming, like any factish, capable of explicating the missteps of an outdated past in which it had "not yet" been taken into account. Or at least capable of convincing us that it is capable. There is no reason to criticize the creation of this ability to convince us, which celebrates the coming into existence of a new being. But this does not mean that we have to let ourselves be persuaded. It is not impossible to celebrate and retain a memory. This is also the meaning of the anamnesis, the voluntary evocation I wish to attempt.

16

Energy Is Conserved!

There are no neutral narratives. A narrative presentation begins—assuming this is the intent—long before it is able to make explicit the perspective it is framing. I have laid out my interests, but the way I have constructed "The Invention of Mechanics: Power and Reason" has already engaged the reader in ways that harmonized with my own perspective. There, I contrasted two of Lagrange's successors: Hamilton, for whom mechanical energy is not just conserved when a mechanical, or dynamic, system undergoes change but becomes, through the Hamiltonian operator, a fulcrum for all the representational changes possible for the system, including the specific change known as movement; and Carnot, the inventor of a very different use of the concept of conservation, which no longer describes autonomous change but corresponds to the ideal of completely controlling the power of heat to produce mechanical work. In the case of the heirs of the Lagrangian event, I did not have to slow down the movement of history. Every physicist is familiar with it. The only difference might be in the insistence of my emphasis. Every physicist "knows" that Hamiltonian dynamics introduces only "conservative" forces that conserve mechanical energy. But some don't see anything remarkable in this restriction. All of Carnot's successors "know" that the change from state to state that constitutes the ideal cycle invented by Carnot, a cycle wherein the equivalence between "cause" and "effect" is ensured *against* the "natural" tendency of heat to pass spontaneously and without mechanical consequence from a hot body to a cold body, is merely a laborious "mimicry" of conservative dynamic change. But the majority accept that this mimicry acknowledges its subordination to the original model. We need to slow down, however, when we are no longer dealing with what customarily "goes without saying" but with the question today associated with the hierarchy of physics. Why do the requirements of dynamics, as I have presented them, no longer impose any limitation on its relevance? Why does the fact that a "force" or a phenomenon is "dissipative" today simply indicate that its definition is "approximate," marked by human finiteness and ignorance, in other words, a part of "phenomenological physics"? What has happened?

The reader knows that the next step will be to bring these two traditions into contact with one another. Maybe she is already anticipating the dramatic turn of events that transformed the perspective of physics in the mid-nineteenth century. From human respiration to the steam engine, from the burning candle to the electrochemical battery: all phenomena, whether part of nature or artificially produced, *conserve energy*. So, I have led the reader to "expect" to see the conservation of energy framed within the context of a very specific problem, one that establishes a very specific connection between the question of the Lagrangian heritage and an event that has affected the very identity of physics, that has transformed the evaluation of what the practice of physicists allows them to claim: the discovery of the conservation of energy.

The conservation of energy is the quintessential example of this type of knowledge, which for Planck established the physicist's vocation: faith in the ability of physics to reveal an intelligible world, independent of our interests and practices. To reveal a world and not construct an objective definition—herein lies the difference between Planck's vision of the world and the object of mechanics that made its way out of Galileo's laboratory and was consecrated by Lagrange's equations. The mechanical object had the power to dictate the way in which it would be defined, and this was the primary reason for its interest. It made it possible to bring together around it those who would invent the mathematical representation it sanctioned but not to gather together disparate phenomena, to represent the world. On the contrary, it had to be selected within that world, then isolated and purified. In other words, it is, with respect to its existence as an experimental object, radically dependent on our interests and practices: the ball has to be round, the inclined plane smooth, and it would be even better if air were absent when trying to satisfy the requirements on which the power of mechanical representation depends. This is not at all the case, however, for the energy conserved—at least after 1850—by any natural process, whether selected by practitioners or part of nature (which was soon to include the stars).

As we now know, the discovery of the conservation of energy was one of those "simultaneous discoveries" that draw the attention of historians of science. As if it were "in the air," so to speak. However, Thomas Kuhn has also shown that statements made after 1840 can only be assimilated retrospectively,[1] each author conferring a distinct meaning on what would become "energy" (and what, at the time, was usually called "force").[2] Contrary to Galilean acceleration, for example, where the inclined plane produced both the measurement and its interpretation, the devices that revealed the conservation of energy were unable to confer a determinate interpretation on their results. A scientist confronting such a result was not forced to see what the author of the device wanted her to see. The matter was open to discussion—of which there were many.

To understand the background of those discussions, we need first to distinguish conversion from conservation. The notion of conversion between "forces" was initially an aesthetic idea, which communicated with the presentation of an "indestructible force" that gave nature its permanent unity. As such, the idea did not, strictly speaking, have an author. We can trace it back to Leibniz's "live force" or to the post-Kantian philosophy of nature. The novelty that characterized the first decades of the nineteenth century is found in the ability to see older phenomena (a burning candle or the heat given off by a chemical reaction) and new (electrolysis, the electric battery, the steam engine) as unanimously confirming universal convertibility. A collection of dispersed "facts" from distinct practices, with distinct interpretations, can be unified if it is seen as a "network" ensuring the conversion of any kind of force (or energy) into another. This was not a "thesis" that would have been negotiated among the different protagonists but a "way of seeing," an aesthetic, that brought together precursors or authors of statements we judge to be "scientific"; and such statements came from physicians, engineers, meteorologists, and physicists specializing in motion, heat, electricity, or magnetism. Weren't they all involved in conversion processes?

The notion of conservation, however, implies measurement. It is not only a question of indestructibility, for what is now at stake is the creation of a device that can be used to *quantify* the conversion. In 1843, Joule identified the quantitative equivalence between heat and work by correlating the rise in the temperature of water in which a system of blades turns to the work needed to produce the movement of those blades. The conversion of mechanical work into heat can now be characterized by a "mechanical" equivalent of heat: this will be the amount of work necessary to increase the temperature of a kilogram of water by one degree.

While the device used to determine the quantity of what

disappeared and the quantity of what appeared doesn't reject the egalitarian network of conversion processes, it certainly distorts it. In fact, measurement privileges mechanical work, which will become the common standard of reference. Correlatively, it privileges laboratory practitioners, for physicians and naturalists are incapable of submitting "their" energies to this type of measurement. But here we must proceed cautiously. Joule's measurement may indeed have taken place in the lab, but it is not, like the measurement of a Galilean body, objective in the strong sense, simultaneously creating the conditions for understanding the phenomenon. It is an engineering measurement, based on the concept of work where, as we saw in "The Invention of Mechanics," the price of generality is silence concerning the nature of what is being measured. Measurement requires that two phenomena be related by an equivalence, yes, but this relation is contingent upon the measurement device, unlike measurements involving the pendulum, where motion is converted "spontaneously" into potential energy and vice versa.

Measurement requires equivalence, but does equivalence provide relevant access to the intelligibility of natural processes? For what may be the first time since Newton's adversaries questioned the breakdown of light by a prism in the eighteenth century,[3] the question of what the laboratory does, of the relevance of the operations it makes possible, has become critical. As Friedrich Engels, a connoisseur in this matter, noted: "If we change heat into mechanical motion or vice versa, is not the quality altered while the quantity remains the same? Quite correct. But it is with change of form of motion as with Heine's vices; anyone can be virtuous by himself, for vices two are always necessary. Change of form of motion is always a process that takes place between at least two bodies, of which one loses a definite quantity of motion of one quality (e.g. heat), while the other gains a corresponding quantity of motion of another quality (mechanical motion, electricity, chemical decomposition)."[4]

In other words, quantitative equivalence cannot be used to contradict qualitative transformation, to reduce it to an underlying identity, because it is associated with a condition—*there must be two,* an interaction must take place—about which it remains silent. The apparatus used to show that what one gains the other loses subjects transformation to the imperative of measurement, but that measurement by itself is incapable of identifying what it makes equivalent. What is heat? How does it differ from mechanical work? How is chemical energy or electrical energy different from that work? The equality reveals nothing. The conflict over interpretation has begun.

The axis of the conflict is obviously the relationship between "mechanical" conservation and the new, "energy" conservation. This gives a central role to the notion of work, the shared reference for the two types of equivalence. Work is "mechanical currency" and, as such, has been used as the common unit of measurement, but, again as such, it is quite incapable of providing the "reason" for the energy transformation, being merely the standard. Does the conservation of energy nevertheless reflect the secret omnipotence of "mechanical reasons," justifying the reduction of qualitatively different forms of energy to mechanical energy alone? This is Helmholtz's thesis. Or, moving to the opposite extreme, does it allow us to question, on behalf of the logic of qualitative multiplicity, mechanical reason itself, that is, the privileged intelligibility of (rational) mechanics? This is Engels's thesis. Following Helmholtz's interpretation, force and work together characterize a world hidden from direct observation. The ideal pendulum triumphs over the imperfect pendulum, whose movement is gradually dampened, for the spontaneous dampening of mechanical motion has as its equivalent the release of heat, which is itself only a form of hidden mechanical motion, possibly analogous to the vibration of atoms in matter. Following Engels, even in the case of mechanics, there must, upon closer inspection, be "two parties": there must be an interaction so that the energy associated with motion is converted into potential energy, and vice versa. For Engels, work is, in all cases, a practical measurement that depends on artificial devices. Thus, the ideal pendulum becomes an unreliable witness in that it appears to justify making the equivalence measured by work the reason for its movement and, therefore, to assign a purely mechanical identity to cause and effect. But Joule's apparatus is now a reliable (modest) witness. It illustrates the instrumental nature of measurement and enables work, as well as the mechanical force corresponding to it, to be interpreted. These are purely operational notions, which are neutral about the identity of the terms whose quantification is made possible by reciprocal measurement.

The practice of the physicist is at stake here. For Engels it entailed a lucidity that questions the very meaning of mechanics: the Galilean object seemed to confer an objective character, dictated by the object, on categories of measurement, but this power appeared to be retroactively contingent, a nonrepresentative particular case of what we can require of nature.[5] For Helmholtz, on the other hand, the conservation of energy justified a universalization of the requirements of mechanics, which no longer defined only the ideal object of mechanics but the conditions of intelligibility of any natural phenomenon. All other fields of inquiry, therefore, had to accept the central importance of the quantitative equality of cause and effect.[6]

In making Helmholtz and Engels symmetrical proponents of the two most antagonistic interpretations of the conservation of energy, we indicate that we are outside the history of physics properly speaking, where this symmetry doesn't exist. In the context of that history, one is a respectable, if mistaken, protagonist, while the other is most often described as an ideological intruder. But this is a hasty judgment and needs to be slowed down to become interesting. A problem then arises. For it indicates that the invention of the "properly physical" issues of conservation is not defined in the arena that circumscribes the tension between Galileo's pendulum and Joule's system of moving blades. In fact, there was another protagonist, and this protagonist descends in a direct line not from the physics of forces but from the rational mechanics of changes of state—that is, the ideally reversible cycle of transformation introduced by Sadi Carnot.

After a complex history, the Carnot cycle would become the "arena" in which the relationship between mechanical energy and what was to be called "thermodynamic" energy, governed by two principles, would be determined. I will discuss that history and those principles in the following pages, but first I want to point out the contrast between the problems they introduced and the related problems the conservation of energy has led to (and which it will continue to engender in "scientific culture").

The conservation of energy introduced "important questions" and was able to attract the interest of a broad range of disciplines: philosophy, physics, biology, medicine, sociology, economy, psychology. Freud comes to mind, but also the physicist Wilhelm Ostwald, who retraced human history in terms of the energy resources made available by human technology and analyzed the psychopathological episodes that characterized the lives of "great men" in terms of energy output. The conservation of energy was a "cultural event" with indeterminate limits, and it is reasonable to assume that the "scientific event" that played out in the arena of the Carnot cycle cannot be disassociated historically from this cultural event any more than Galileo's laws can be disassociated from his confrontation with the Catholic church. But, like Galileo's laboratory, the arena defined by questions about the Carnot cycle was unique in the sense that the issues that arose there could only be understood and discussed by specialists. It is not so much a question of competence, even if the formulation of the issues makes it necessary to grasp the difference between energy transformations that can be associated with a series of changes of state and ordinary energy transformations, for example, those involved in Joule's system of blades.[7] Aside from competence, it is interest that drives the selection of the protagonists. Neither the biologist, nor the physician, nor for that matter the dialectical philosopher, has any reason to take an interest in the now crucial issue constituted by the concept of change of state. Nothing they are involved in adds any relevance to the concept.

Moreover, it is not simply a question of understanding how the Carnot cycle came to serve as the arena in which what we respectively refer to as the "cultural" and "properly scientific" issues of the conservation of energy were differentiated. We also need to understand its somewhat strange status within twentieth-century physics. The first-year student of physics or chemistry still learns the Carnot cycle in the way Clausius redefined it, but this "required chapter" in the curriculum most often inspires boredom and confusion. The student doesn't really understand why it is even necessary. Therefore, the arena did not have the ability to define its issues but only to serve as a framework in which questions whose answers were found elsewhere were formulated.

In any event, we can understand the student's puzzlement: the Carnot cycle reinterpreted by Clausius has become a very strange creature. Its invention by Carnot was part and parcel of a science whose demise the conservation of energy announced: the science of heat was identified with a fluid that was conserved and whose behavior could be used, in particular, to explain the experimental relationship between pressure, volume, and temperature that characterizes a gas.[8] The science of caloric was a cutting-edge science in the first half of the nineteenth century and it is this that Carnot not only worked with but connected with the great mechanical tradition of the conservation of a cause in the effect. This connection, invented by Carnot and realized in his cycle, lost its justification with the destruction of caloric heat. The great irony of the Carnot cycle, the one that often provokes the disgust of students but that also serves as the arena in which the "before" and "after" of the conservation of energy are measured, is that the cycle itself and the optimal yield it defines have survived. It is a connection that no longer connects, a bridge that ties together two banks whose highway systems have been modified so extensively that one has to wonder why it was built in the first place. It is as if the Carnot cycle, which captured energy transformations, was an abstraction that came from nowhere and to which no intuition leads.

The cycle appears as just such an abstraction because the caloric theory, that is, the conservation of a "heat-substance," which led to its invention and made it intuitively intelligible, is now forgotten. The amount of caloric contained in a body is obviously not directly measurable according to the terms of this theory. But one thing is certain: if a given quantity of gas, whose volume and pressure have changed, and which has received or given off heat, returns to the initial values of its pressure, volume, and temperature variables, it means the gas must have given up as much heat as it took in during the cycle of its transformations. That is why Carnot was able to define the states of his cycle in terms of pressure, volume, and temperature without having any way to determine how much heat was absorbed from the hot source and how much was transferred to the cold source. Regardless of the path taken in moving from one state to another, whether the system is compressed at constant temperature and then cooled or cooled at constant volume and then compressed, the conservation of caloric ensured that, from the moment the cycle becomes closed, all the heat absorbed has been restored. And this holds true even for "real," that is, nonideal, cycles. Also, any transition from one state to another, where both are characterized in terms of pressure, temperature, and volume, should imply that, regardless of the path between those two states, the same quantity of caloric is absorbed or released, the only difference between the paths—or between ideal and nonideal transitions—being the mechanical work produced or consumed at the time. In other words, the conservation of heat or caloric served as a fixed point, ensuring that the "same quantity" could be described, even if this quantity could not be measured. When the caloric theory gave way to the conservation of energy, when Carnot's apparatus stopped transmitting heat from a hot source to a cold source but converted the heat into work, the cycle no longer offered any obvious guarantee about anything. Quite the contrary, it now became a problem: why couldn't all the heat received from the hot source be converted into work?

The same holds true on the other "bank," where we find the conservation of cause and effect. Carnot had shown that because of its reversibility, the output of his ideal cycle was better than any other device could provide: for a given amount of caloric passing from a source of heat to a source of cold, it produced the maximum possible amount of mechanical work. However, Carnot's demonstration was based on a reductio ad absurdum that was traditional in mechanics. For if a hypothetical cycle had greater output, then by coupling it to an ideal Carnot cycle operating in reverse, as a heat pump, it would *freely* produce mechanical work. But if the heat is converted into work, the absurdity disappears: the result of the coupling is not the free production of work but the conversion of more heat into work. Once again we are forced to ask, why can't all the heat be converted into work? The optimal output defined by the ideal Carnot cycle has become an enigma.

If the Carnot cycle, part of a body of physics that died with the arrival of the conservation of energy, has survived, it is because it identified and implemented an *ideal operation,* whose reversibility in itself guaranteed that all loss had been eliminated. But the loss of what? That is the question we must now answer. Not energy, for this was both a certainty and the primary difficulty. Whether heat flows directly between two bodies at different temperatures or is "converted" into work, measured by a change in volume, *in all cases energy is conserved.* Because the cycle is reversible, it introduces a cause that is conserved in the effect it produces, but this "cause" must be completely distinct from energy, for the energy balance is completely indifferent to the ideal, conservative nature of the cycle. The cycle speaks of the impossibility of converting, for a given temperature difference, more heat than is defined by Carnot's optimal output, but the energy being conserved is silent about the question of possible and impossible conversions between distinct forms of energy. In other words, the conservation of energy doesn't have the ability to characterize the ideal invented by Carnot because it is indifferent to the reversibility of the cycle, just as it is indifferent to the distinction between the ideal pendulum and the real pendulum whose movement is slowed down by friction. So, in relation to what conservation is the loss eliminated by the cycle defined?

The challenge of defining what it is that reversible transformations conserve may well be what separated the new "scientific" thermodynamics from the "great questions" brought about by the conservation of energy. We know, from the history of mechanics, why reversibility is privileged. Mechanical movements that conserve energy, that do not gradually slow down as a result of friction, can be described by a state function, that is, expressed as a series of changes of state. And it is this possibility that has given mechanics its formidable inventive power and, at the same time, limited that power to the class of ideal movements alone, devoid of friction. The general conservation of energy seemed to have flattened this difference. Because a movement that is slowed by friction "conserves" energy, a part of the mechanical energy is "simply" converted into heat. But the energy that is conserved, precisely because it is always conserved, has lost its status as a state function. It can no longer distinguish between the ideal situation, where the "cause" is conserved in its "effect," and the dissipative situation, where the cause is exhausted in producing a lesser and sometimes null effect (as is the case when heat is spontaneously transmitted from a hot body to a cold). This is the new message of the Carnot cycle: even if all energy transformations conserve energy, they are not all the same, and it is the definition of this nonequivalence that needs to be made explicit. In other words, the ideal output defined by Carnot, the determinate character of the relationships between heat energy consumed and mechanical energy produced at the conclusion of the ideal cycle, traces the enigmatic figure of a new *state function*.

17

The Not So Profound Mystery of Entropy

The following analysis is obviously retrospective in the sense that, in order for the problem to be formulated, the output defined by Carnot had to find a defender capable of saving it from its association with the caloric theory, a physicist willing to risk creating the problem of the meaning of the reversible cycle in a world where energy is conserved. This role was given to William Thomson, the future Lord Kelvin. Starting in 1847, Thomson tried to help his fellow Englishmen understand the operation of their steam engines from the point of view of the ideal system defined by Carnot. In January 1850 he verified an unforeseen experimental consequence of Carnot's theory (the freezing point of water drops when the pressure increases). But it is also in 1847, a time when the Carnot cycle defined the scope of Thomson's research and expectations, that he heard Joule describe his experiments for the first time: heat, far from being conserved, could be produced by mechanical motion. For the follower of Carnot, this was the beginning of a long nightmare. In an article from 1849, Thomson compared Carnot to Joule: Joule claimed that nothing is lost in nature, that energy is never destroyed, and yet Carnot's ideal output implies that, when heat flows directly from a warm body to a cold body, the mechanical effect it might have produced is lost. In such cases, what else is produced in place of that lost effect? Thomson concluded that there could be no theory of heat until the question was answered.

It was a little-known physicist, Rudolf Clausius, who, in February 1850, would provide an answer of dizzying simplicity. Clausius hadn't read Carnot (whose paper was unavailable), but he had read Clapeyron, Carnot's only French disciple, and Thomson, and succeeded in cutting the Gordian knot with a single blow: "I don't consider the difficulties to be as great as Thomson thinks," he wrote.[1] In fact, Clausius would show that all that was required was to abandon one of Carnot's axioms, the one that states that, at the end of the cycle, all the heat removed from the hot source has been restored to the cold source. We need to "see" the cycle as carrying out two simultaneous operations: *conversion* of part of the heat taken from the hot source into mechanical motion, and *transmission* of the remaining heat to the cold source.[2] Carnot's ideal output established the maximum ratio between conversion and transmission. For a given quantity of mechanical motion produced, any nonideal cycle transmits to the cold source a *greater* quantity of heat than is ideally possible.

There is no need to point out that Thomson got no pleasure from seeing himself overtaken in this way. In 1851 he published his own analysis of the Carnot cycle. A quarrel about priority ensued, which I won't pursue here as I want to resist the temptation to "do history." I introduced the two rival readers of Carnot—Clausius and Thomson—because the difference between them clarifies a problem we have inherited. For many people, anyone who speaks of the Carnot cycle speaks of "irreversibility." We owe this association to Thomson, although it was somewhat misleading in that the cycle had, on the contrary, created the means to refer to reversible energy transformations. And it is Thomson who, in assuming that energy conservation was to be understood as a transformation between "cause" and "effect," wondered what it was that heat "caused" when the thing the ideal Carnot cycle was designed to eliminate was produced, namely, the direct flow of heat between two bodies at different temperatures. For Thomson, the Carnot cycle highlighted the question of the loss whose elimination defined the ideal operation of that cycle: what was lost in a world where nothing is lost? As for Clausius, he would adopt the position that Carnot, the heir of rational mechanics, the creator of a device for producing equalities, had assumed.

The distinction between the two readings became apparent as soon as a response to the crucial question was required. What would become of Carnot's reductio ad absurdum regarding the optimal output of his cycle? What absurdity would describe a machine whose output was higher than Carnot's ideal output? In other words, what would such a machine do when coupled with Carnot's ideal machine running in reverse? The two readings of Carnot are apparently in agreement. The coupled machines would be able, with no need of mechanical input, to transfer heat from the cold source to the hot source, a possibility quite compatible with the conservation of energy. The new reading of Clausius and Thomson received significantly different formulations. For Clausius, a heat "pump" that freely generated a thermal difference was impossible, for it contradicted the "essence of heat, which always tends to equilibrate existing temperature differences." For his part, Thomson tried to give this impossibility a form that would parallel the statement he proposed to associate with the conservation of energy. This statement of conservation reflected the impossibility of a "perpetual motion machine of the first kind": no device can freely create work or motion, that is, energy. The ideal Carnot cycle stated the impossibility of "perpetual motion of the second kind": no device can extract work from the heat in the surrounding environment without simultaneously transferring part of that heat to a colder body. Thomson's ambition was to construct a "thermodynamics," a dynamic science of heat, from two principles that were as symmetrical as possible. This project would be made explicit in 1852 when he succeeded in giving "Carnot's principle" a formulation that eliminated devices, motors, and perpetual motion. Just as the conservation of energy can apply to any situation, we can extend the lesson of the cycle to any process, natural or artificial. But it is the "loss" that has now become the main subject of this lesson, for it is in terms of loss that the ideal cycle defines the "natural" processes it manages to avoid. Every nonideal cycle and, in general, any energy transformation results in dissipation without the return of energy. Part of the energy that, before this transformation, *might have been* converted into mechanical energy no longer can be. In this way, a possible effect has been irreversibly lost. For Thomson, the degradation of energy is a principle as universal as that of conservation. It affirms the gradual disappearance over time of all "usable" energy, that is, energy capable of producing work, and, eventually, the "heat death of the universe," the final state to which the universe tends when all its energy has become "unavailable."

When we think of the "degradation" of energy, we think of "entropy." And certainly, from the point of view of imaginative rhetoric and the fascination exerted by entropy, we are Thomson's heirs. But if entropy is so mysterious, so hard to explain, it is precisely because, as a concept defined by Clausius only in 1865, it was not intended to define "energy degradation," or characterize natural processes as a whole. And although it is not incapable of describing the "loss" of available energy and "irreversibility," it does so in a peculiar fashion, without ever abandoning its reference to the ideal defined by Carnot.

The major difference between Thomson and Clausius is that Clausius never looked for symmetry between the principle of energy conservation and the "Carnot principle." And when he provided a "realistic" and symmetrical formulation of the two principles, he did so by introducing cosmological limitations: "The energy of the universe remains constant. The entropy of the universe increases toward a maximum," wrote Clausius in 1865. Those two statements are marked not only by their great sobriety but by an unusual sense of irony. For, in this case, the universe is not the symbol of the "absolute" character of the law of increasing entropy. For Clausius, it is, rather, the only "system" that, by definition, does not undergo exchange with an environment. That is why it is the only case that allows energy and entropy to be subject to statements of similar scope.

Note that if Einstein's general theory of relativity gave pride of place to the universe, it wasn't because physical problems were resolved everywhere else but because a homogeneous and isotropic universe is one of the rare objects simple enough to be explicitly treated using Einstein's equations. Similarly, Clausius's cosmological formulation does not imply that because all natural processes result in an increase of entropy or degradation of energy, the conclusion can be extended to the entire universe. Quite the contrary, the universe is the only case where the difficulty concealed by the apparent generality of Thomson's "degradation of energy" doesn't arise: although energy is always and everywhere conserved, the extent to which energy is or is not degraded, the extent to which it is or is not usable, depends on the circumstances. Thus, the radiation emitted by the Sun is irreversibly "dissipated" from the point of view of the Sun but nonetheless "used" on Earth by every living thing capable of "exploiting" it. Engels's qualitative diversity returns in the diversity of conversion devices. In other words, we are not allowed to forget devices when dealing with natural processes. And Clausius knew this. As we will see, he would show (contradicting Thomson) that only Carnot's ideal cycle can assign a "measurable" character to the loss or "degradation." It can only be evaluated relative to the ideal reversible transformation, that is, relative to a human device.[3]

The relative merits of Clausius and Thomson are not the issue; what I would like to do is clarify what it is that makes entropy so often perplexing. Why does a concept of such quasi-prophetic allure refer us back to some antiquated device, which seems part and parcel of the now humble problem of the output of steam engines? And the first thing to understand is that, in Clausius's hands, the Carnot cycle has little in common with the ideal model of the steam engine. Clausius, by way of Carnot, is the heir of Lagrange and the great tradition of mechanics, where equality is the fulcrum providing the freedom to transform definitions and construct fictions that reflect the singularity of the object. It wasn't the loss of available energy or the system's output that initially interested Clausius but the reversibility of the cycle. That reversibility constitutes the very heart of the argument leading to the definition of what is conserved in Carnot's ideal cycle.

Therefore, it's the cycle as a whole rather than any particular energy transformation, ideal or otherwise, that will be fictionalized. Clausius asked that we stop asking "local" questions. At what stage does a conversion occur? At what stage does the flow of heat take place? He used the ideal cycle because it allowed him to write an = sign between *a* flow and *a* conversion. When the cycle is ideal, it is the locus of two transformations that exactly balance one another. If such an ideal cycle has absorbed a quantity of heat, Q_1, at temperature T_1 from the hot source, and has given up a quantity of heat, Q_2, to the cold source at temperature T_2, we can and should say that the flow of heat Q_2 between temperature T_1 and temperature T_2 exactly compensates for the conversion into work of the amount of heat represented by $Q_1 - Q_2$. The ideal cycle allows us to introduce an equivalence value that defines the "exact price" of a conversion in terms of heat flow.

There are two ways in which Clausius's treatment follows in a direct line from the heritage of Lagrange. First, the ideal is dramatically separated from the real. Contrary to the equivalence between heat and work measured by Joule, Clausius's equivalence value does not correspond to the evaluation of a specific, concrete process, it is justified by the ideal, reversible nature of the cycle. Correlatively, and this is the second point, the equivalence value may appear to establish a *price,* but it's the possibility of establishing a price, a compensation, by means of the ideal cycle, that defines the corresponding terms—what pays and what is paid for. In short, it's the equivalence that controls and distributes their identity to the terms it articulates. The ideal has become the only true subject of the description in that it authorizes the fictionalization of the cycle, its transformation into an equivalence-producing operator.

We shouldn't be surprised that the next step, for Clausius, was the definition of a *state function.* The goal was to make the transition from the "exact price" of conversion to a definition that harmonized with mechanical energy in the Lagrangian sense, namely, the definition of the "price" of the transition between any two states of the cycle, independently of the path taken, as long as the transition is ideal, that is, reversible.[4] To construct this state function, Clausius would transform the Carnot cycle into a purely fictional device. This involved making full use of equivalence, the = sign between flow and conversion that allowed him to describe the closed cycle. For example, Clausius imagined a cycle with three sources. The system is not realistic, but it allowed him to state the equivalence in algebraic terms, whereby a quantity—here the equivalence value of the conversion of a given amount of heat into mechanical energy—is defined in terms of other quantities, the respective equivalence values of two heat exchanges, each at a given temperature.[5] This allowed him to construct an equality that could be used to measure all transformations on the basis of a single "currency": the exchange of heat at a given temperature. The state function defined by Clausius is the triumph of a rational fiction, of an "as if" justified by compensation.

With the conclusion of his work in 1854, Clausius had

defined the state function that characterizes the ideal cycle: Q/T. For any reversible transformation between two states, regardless of the path, there can be a corresponding "distance": the amount of heat exchanged divided by the temperature of the exchange. Take, for example, two states separated by an infinitesimal distance dQ/T, which is the "cost" of the transformation that leads from one to the other. What does this sort of evaluation imply? It implies that, regardless of the infinitesimal (reversible) path between the two states, it is equivalent to a particular path, where dQ/T has a clearly determined meaning. This path consists of an infinitesimal fragment of an isotherm (where the system receives a quantity of heat, dQ, at temperature T), and an infinitesimal adiabatic fragment, where the system is thermally isolated ($dQ = 0$), but temperature varies. Clausius's definition generalizes Carnot's invention: a cycle whose very conception ensures its reversibility. This generalization meant representing a given path as a succession of particular fragments, where each fragment is characterized by a change imposed on a controlled variable (Q or T) while the second (T or Q) is held constant. We should try to imagine the space that Clausius invented to represent the distance between any two states of a system as defined by the dense set of all the isothermal and adiabatic curves characterizing this system. This set forms a system of meshes that can be as fine as we wish, in terms of which all reversible paths imaginable between two states of this space can be redefined. Consequently, any path can be represented as an infinitesimal succession of isothermal (dQ/T is determined) and adiabatic curves ($dQ/T = 0$), similar to those that make up the four stages of the Carnot cycle. The equivalence value of the transition between any two given states will then be the integral of dQ/T corresponding to the succession of isotherms along the transition.[6] By definition, the integral of dQ/T for an ideal cycle is equal to zero. This is then the sought-for state function, exhibiting what Carnot's ideal cycle conserves.[7] The first definition of what will become entropy, as a state function corresponding to all the various energy transformations, is, in the case of Carnot's ideal cycle, that it has the value Q/T.

I have discussed the state function defined by Clausius at some length for two reasons. The first is, in a sense, cultural. Today, it is seen as good form to assimilate manipulation and measurement to a utilitarian concern that could be contrasted with the "noble" activity of meaning creation. Yet, Carnot's work, completed by Clausius, is entirely focused on the manipulation, control, and production of measurement, although, at the same time, it also demonstrates the invention, the free and entirely counterintuitive production of meaning implied by the creation of certain types of measurements. We can, of course, measure anything, unilaterally decide to use the same type of measurement for the activity of laborers and the motion of the Galilean ball. But we then have no way of establishing a relationship between the measurement and "what" is being measured. In the case of the ball and other mechanical objects, this relationship appears, on the contrary, fully determined: the mechanical object is defined as measurable, defined by the equivalence used for the measurement. In the case of energy transformations, however, measurability is in no way a "given," it must be created, fabricated from whole cloth; in this case, the definition of the ability to measure something is not arbitrary, it creates the very object it measures.

Correlatively, the measurement defined by Clausius entails requirements and obligations. And this is precisely the difference between Clausius and Thomson. For both men, not all energy transformations are equal, and the "second law of thermodynamics" makes nonequivalence explicit. Thomson, however, sought to apply this nonequivalence to processes themselves, as was the case in mechanics. As with conservation, the degradation of energy was supposed to characterize processes "in themselves." Nonequivalence was supposed to be "objective" in the sense that it didn't obligate the physicist to anything in particular, being "dictated" by phenomena. Clausius, for his part, reinvented the Lagrangian tradition transmitted by Carnot by clarifying the requirements and obligations of *rational* measurement, justified by a state function and, therefore, the power of the = sign. If all energy transformations are not equal, it is because only reversible transformations satisfy the requirements needed to define the appropriate state function. Rational measurement requires the reversible ideal. And—and this is the new factor differentiating mechanics from thermodynamics—it obligated the physicist to be conscious that he was a manipulator, an active participant in the definition of equivalence. The "change of state" measured by Clausius has nothing to do with the spontaneous transformations produced in nature. On the contrary, it implies that all spontaneous "natural" evolutions have been eliminated. Whereas the ideal is defined by the absurdity of a machine whose operation would produce a gratuitous increase of temperature differences, thereby confirming a world in which temperature differences are spontaneously equalized, this leveling off, like any spontaneous change, cannot be described. The description takes as its only objects transformations driven by outside manipulation, pseudo-changes wherein the system is in fact constrained by the manipulator to transition from one equilibrium state to another that is infinitely near.[8]

If, as Kant claimed, the Copernican revolution marks the point where scientists now ask questions, and subjects phenomena to their categories, there is, in the case of thermodynamics, no profound mystery about such submission: the Copernican judge needs hands, he has to fabricate, here pilot, the subjected "object." The reversible transformation is a human artifact and its artificial character has nothing to do with purification (smoothing the inclined plane, polishing the billiard balls, traveling to the Moon, where there is no air) and everything to do with creation.

The second reason has to do with the confrontation for

which the "Carnot cycle" will be the arena, and entropy the prize. We are not quite there yet, or rather, and this is the interesting point, we are not at all there. If the Carnot–Clausius reversible transformations "mimic" dynamic changes and are unable to designate a "natural referent," like the fall of a presumably ideally smooth billiard ball, how was the cycle able to accommodate the confrontation between the followers of Carnot and those of Hamilton? Maxwell, it should be pointed out, did not ridicule the Carnot cycle, which fascinated him; he laughed at the claim of German physicists that a relationship of some sort existed between Hamiltonian dynamics and the laborious mimicry that artificially reproduced some of its incommunicable attributes. And so the question can now be asked: how can entropy, a state function that ignores the time required for spontaneous changes because its definition involves transformations that are fully controlled by human operations, support such a claim?

In the general definition Clausius would give it in 1865, entropy (from the Greek τροπη, or transformation) has become the appropriate state function for any cycle, ideal or not. This means that in any transformation cycle, ideal or not, as long as it is complete, that is, as long as the "body" returns to its initial state, the integral of the infinitesimal changes in entropy, written dS, is by definition identically zero. Of course, in all nonideal cases, the change in entropy, dS, ceases to be equal to dQ/T, the state function for reversible transformations. This may not appear to be unduly serious but it has drastic consequences. When the cycle is not ideal but includes spontaneous transformations that are identified (from the Carnot–Clausius perspective) with losses, dQ/T is no longer a state function, although entropy is still referred to as such. However, it is now nothing but a name, for in the nonideal case there is no determined relationship between entropy and the variables that characterize the system.

For general situations, we can certainly write $dS = dQ/T +$

dQ'/T. But the amount of heat Q is now restricted to describing exchanges that are "compensated for" by the work that is actually produced, while Q', the "uncompensated" heat, refers to the heat that has been wasted, without compensation. And this is all we know about it because measurement of the "loss," the heat that was irreversibly wasted, does not characterize the real cycle as such but only when contrasted with the ideal cycle. To measure it, we simply need to couple the nonideal cycle to an ideal cycle running in reverse, that is, one that uses the work produced by the first: the amount of heat that flows to the cold source during the nonideal cycle but which the second (ideal) cycle is *unable* to return to the hot source is the uncompensated heat.

In other words, the "entropy" state function does not allow us to escape the ideal invented by Carnot. It does not allow us to define irreversible processes other than as "losses" with respect to the reference ideal constituted by the reversible transformation. No physical interpretation will allow us to escape the reference to the ideal compensation mobilized by the measurement of loss, that is, the quantity of heat the coupled cycles have been unable to return to the hot source from the cold source. Because the coupled cycles leave behind an amount of heat at the cold source, they do not restore the initial state, and the final entropy is not equal to the initial entropy. The = sign, whose power the state function continues to express, is powerless to identify what it relates.

So, even though entropy may be a generalized state function, conserved through any cycle, ideal or not, *its definition does not confer any power on the physicist* once the cycle is no longer ideal. More precisely, the only power the physicist can claim is the power to define the sign of uncompensated heat, $dQ' > 0$, and this power reflects what everyone knows: that uncompensated heat corresponds to a loss.[9] The reverse case, where compensated heat is negative, would correspond to the "absurdity" of perpetual motion of the second kind, the free increase of temperature differences. But the loss can only be established or evaluated with respect to the ideal cycle, and is not connected to any description, realistic or fictional, of the processes responsible for its production. Therefore, the fact that irreversible energy transformations always result in an increase in entropy is simply another way of saying that they are always defined as a loss with respect to the reversible ideal.

Entropy, then, is a rather strange state function. It appears to subject every energy transformation to the "rational" logic of state functions, but doesn't correspond to any definition, or any systematic relationship among measurable variables, this relationship being limited only to those cases where the transformation is reversible. Through this strange state function the time taken by "natural" processes appears to gain a foothold in thermodynamics—the fact that any irreversible transformation increases entropy apparently refers to an increase over time—but such an increase is in fact clearly defined only in ideal situations, where the time relative to the manipulation of the system completely replaces the time it takes for a process to occur. And this is, certainly, the end result Clausius wished to obtain from entropy as he understood it: to silence his rival Thomson and demonstrate that the degradation of energy over time does not generally have any well-defined physical meaning.

And yet, entropy has continued to fascinate us by its message of fatality, by its association with loaded terms: degradation, heat death, the arrow of time. But this was not simply a question of "cultural misunderstanding," a synonym of the confusion of public opinion; physicists would presumably have understood Clausius's lesson and accepted the austere limitations corresponding to the requirements and obligations of their practice. For physics has a history of invention, and the invention of new questions always entails risk, and is never a logical or ethical operation. In this case, physicists were themselves subjected to the "mystery" of entropy. Even though it is silent about the nature of the irreversibility whose results it can account for only in terms of loss, entropy nonetheless "represented" for them the problem, or more specifically the challenge, of "irreversible" processes. Isn't it possible to escape the Carnot cycle, and an ideal that can't be attributed to any process, and address "human interests" that are directed toward the evaluation of output and loss? Can't irreversibility, or the increase of entropy, be given a positive meaning? It is at this point that, for some proponents of the question, the requirements and obligations of the Lagrangian heritage change their meaning and become synonymous with the subjugation of physics (thermodynamics) to merely utilitarian interests, to a concern with output. This was a new situation, one that contrasted "practical concerns" with loss or waste with the true concerns of the physicist, who intends to answer the questions presented by a world unconcerned with waste.

However, for this to occur we once again need to hear from the Carnot cycle. Before physicists could determine if and how it was possible to escape the cycle, they had to understand the operations of this cycle in new terms, independent of any reference to "conversion." More specifically, such reference had to recede into the background, becoming a simple consequence of another way of defining. Therefore, whatever it was that made the cycle unique had to be made explicit in a different way. In this case, the question of the ability to give a positive meaning to the increase in entropy corresponds to the introduction of a new actor, which will now occupy center stage: *thermodynamic equilibrium.*

What, in fact, makes up the Carnot cycle? A succession of states of equilibrium. Each of the states through which the cycle passes would remain unchanged if left alone. It is the manipulation that "forces" the system to change its state. And what is a "loss" in the case of the Carnot cycle? Some overly forceful manipulation moves the system a finite distance from equilibrium, and the system undergoes a spontaneous change restoring it to equilibrium. The entropy increases every time a system returns to its equilibrium state as a result of a *spontaneous, irreversible change.* Interest could now be focused on the "equilibrium states" the Carnot cycle passes through. Each of these can be defined as the conclusion of an irreversible change accompanied by an increase in entropy.

Until the last decades of the nineteenth century, the difference between the equilibrium state achieved during a chemical reaction or by a gas whose temperature becomes uniform, on the one hand, and a pendulum, on the other, had not been considered a problem. Of course, everyone knew that if the motion of a pendulum slowed down until it reached a point of motionless equilibrium, it was the result of friction: when the motion of the pendulum is ideal, its equilibrium state is a dynamic state similar to all the others, and simply corresponds to a state of minimum potential energy. The ideal pendulum periodically passes through this state, as it does all the others, without stopping there. The only thing to note about this state compared to the other states along its trajectory is that this is the only state in which a pendulum at rest will remain at rest. Everyone knew that, on the contrary, the leveling off of temperature or a chemical reaction are processes that lead to equilibrium monotonically, equilibrium being the end point of a unidirectional change. However, this evident difference entailed no obligation. When Carnot spoke of "reestablishing the equilibrium of the caloric," he did not feel it was worthwhile to point out the difference from a mechanical state of equilibrium. And it took a certain amount of time before other scientists felt bound by the obligation, which was an obstacle to the direct application of mechanics to thermodynamics. In the early 1870s, the young physicist Max Planck would use it as a weapon against the older Wilhelm Ostwald, whose "energistic" doctrine obscured the difference between the equilibrium of the pendulum and the equilibrium of heat. Even as late as 1855, the young Pierre Duhem's doctoral dissertation was rejected because in it he questioned his senior, Marcellin Berthelot, whose "thermochemistry" implied the equivalence of the two types of equilibrium.[10]

The possibility of using the "second law," the increase in entropy defined by Clausius, to describe the singularity of the thermodynamic equilibrium state was not an individual discovery. Over the course of several years, Massieu (1869), Planck (1869), Gibbs (1876), and Helmholtz (1882) defined different functions for different types of thermodynamic systems, each of which emphasized the role of the second law of thermodynamics in the definition of the equilibrium state of the corresponding system.[11] In 1866, Pierre Duhem referred to all these functions as "thermodynamic potentials." The thermodynamic equilibrium state corresponds to the extreme value of the potential representing the system (entropy in the case of a thermally isolated system). Equilibrium is defined by the fact that the second law of thermodynamics forbids spontaneous changes—changes that are not imposed by manipulating the system—which would cause the system to leave the state characterized by this extreme value. For instance, any spontaneous change that moves a thermally isolated system away from the state defined by maximum entropy would result in a negative value for dQ', and is therefore forbidden. This means that the lesson of the Carnot cycle as something wholly manipulated has changed its meaning. It no longer satisfies the desire to avoid spontaneous processes, synonymous with loss. It results directly from the fact that individual states participating in the cycle cannot modify themselves but require external manipulation.

Once again, this does not miraculously enlighten us. The definition of the state of equilibrium differs profoundly depending on whether we are speaking of mechanics or thermodynamics. The mechanical state of equilibrium is defined by a minimum of potential energy, but every dynamic state can also be characterized by a determinate value of that potential energy and no dynamic state is privileged. On the other hand, we cannot generally characterize a given thermodynamic situation by a corresponding value of its thermodynamic potential. Only the *extreme* case of potential, characterizing the equilibrium state, is defined. Therefore, only the equilibrium state corresponds to a state in the proper sense of the term, namely, one that is characterized, by means of the corresponding potential, in terms of the variables (pressure, temperature, etc.) that define the system. The increase in entropy during an irreversible change toward equilibrium (and more generally the change in thermodynamic potential between a nonequilibrium initial state and the final state of equilibrium) can no more be measured than Clausius's entropy. Only the sign of the change is defined.

No matter. An almost aesthetic transition has taken place. The Carnot cycle no longer mimics Lagrangian trajectories. It has become a device through which thermodynamic equilibrium states have acquired the ability to confirm their difference with mechanical equilibrium states. And it is as such that it can now serve as the arena in which the significance of this difference can be played out. The second law no longer defines the optimal conversion of heat into work, it gives natural processes of energy transformation a unity that is simultaneously very close to mechanics and radically different. The irreversible increase of entropy no longer represents the fact that natural processes can't be made dynamically equivalent without manipulation, it is now seen *as if* it were "positively" describing the contrast between those natural processes and dynamic changes.

How are we to understand an "irreversible" change? How should we interpret the increase in entropy? Such questions are, as I hope I have shown, relative to an authentic history rather than the logical development of a problem that would have "resulted" from the first unification of natural processes under the umbrella of the conservation of energy. More specifically, such questions point out what can be called a "capture operation." As we saw earlier, Engels hoped that the conservation of energy would introduce a crisis into physics, which refused to consider the operation of measurement on which it depended, and would force it to confront the qualitative difference among various forms of "motion." My comments here have shown if not why, at least how this question as such failed to become a historical subject for physics. For the diversity was "captured," "unified" by a common trait that does not refer to it as such but introduces the contrast between all the various processes in which energy, on the one hand, and mechanics, on the other, are transformed. It is in this sense that the Carnot cycle is an arena, a place where actors who might have encountered one another under different circumstances are sworn to combat, that is, must define themselves according to one, and only one, distinguishing feature. What is energy? What does the diversity of forms signify? Once captured, these questions are reduced to the contrast between mechanical and thermodynamic states of equilibrium.

But capture always implies the possibility of "reciprocal capture," the correlative coinvention of two mutually referring identities. What physicist will the second law give birth to? How will she be able to determine what she requires of the "irreversible processes" to which her practice is now addressed, the kind of processes that force her to confront a dilemma: either she subjects them to thermodynamic measurement, thereby eliminating the irreversibility that singularizes them, or she treats them as irreversible, but can then describe them only from the point of view of the equilibrium state to which they lead under certain conditions? Can she look forward to a physical interpretation of the increase of entropy, a physical description of the distinction between a situation of nonequilibrium and an equilibrium state? Can she require of the nonequilibrium situation that it be defined as a state? Or should she say good-bye to that requirement and celebrate the austere rationality of a practice said to have abandoned the realist ambition mechanics was able to nourish and accept with lucidity the limitations to which it is subject? Which of these now divergent values will she advance: realism or a construction that celebrates the singularity of cases wherein description and reason coincide?

18

The Obligations of the Physicist

So far I have introduced an "arena" in terms of which the challenges of extending the concept of conservation to all natural processes have been defined, which is to say, invented. My claim has been that the history that would be played out through them could be read as an operation of capture, the (unintentional) effect of which is a differentiation between true science and "ideology," a differentiation we now take for granted. Although the "great problems" raised by the qualitative diversity of energy forms had been of interest to many, it takes a physicist to be concerned about the difference between a mechanical state of equilibrium, characterized by a potential energy minimum, and a thermodynamic state of equilibrium, characterized by an extreme of thermodynamic potential. I haven't discussed the protagonists who were about to confront one another, only their physical–mathematical representations. Have I fallen into the trap of a "history of concepts," one that represents them as pure creations, detached from the practices and stories of their creators?[1] On the contrary, I wanted to present—to make present—the beings who singularized the history of physics and who, far from allowing it to be explained or summarized, transformed it into a real history, with the kind of suspense and drama that characterize such histories. For the beings I described or am about to describe—the entropy resulting from the Carnot cycle, states of thermodynamic equilibrium, Maxwell's demon, the ambassador of the Hamiltonian Queen of Heaven, and someday soon quantum indeterminacy, and the symmetry difference between gravitational interaction and the three other forms of interaction—are certainly creatures of human history, but they are highly singular creatures, who haunt their creators, and who are given the power to impose their own questions upon them.

In Book I, I introduced the concept of "experimental factishes," such as the neutrino or the Pasteurian microorganism. In "The Invention of Mechanics," other, very different, factishes were introduced, "physical–mathematical" factishes, the mechanical equations that translate and actualize the power and freedom the = sign confers upon the physicist. Now we are faced with "enigmatic factishes," whose singularity resides in their power to impose upon physicists questions we might be tempted to describe as "illegitimate," questions they are not equipped to answer and that any sober analysis of the constraints bearing upon the construction of their objects should exclude.

Illegitimate does not mean irrational but merely reflects the fact that the values activated by these factishes and that cause them to exist, are not the values of proofs and tests but the values of vocation: they are enigmatic in the sense that, without being able to indicate a path, they make reference to another physics and another world. Naturally, enigmatic factishes would be inconceivable without the experimental process or physical-mathematical proof. But they cannot be reduced to either one, although they borrow a particular feature from each. Like the experimental factish, but unlike the physical–mathematical factish, the enigmatic factish introduces the problem of its power in the face of an a priori heterogeneous world. But there is a difference. Experimental devices will be treated as black boxes as soon as they have overcome controversy, will extend their power by multiplying and diversifying those who refer to them, and will be transformed accordingly, until they finally incorporate the existence of users who are meant to follow the instructions of a commercial device. As they extend their power, enigmatic factishes retain a transparency that no practical redefinition can obscure. Like physical–mathematical factishes, they are addressed only to well-defined users, and these, once mobilized, are bound to inhabit the same world, a world defined by the enigma in question, by the proposed vocation. However, these factishes are not models, in the sense that the term implies the power of the imagination to create or identify relationships of resemblance; rather, they are vectors of obligation, conferring upon the questions they allow to be asked the power to engage, authorize, or prohibit.

The appearance of "enigmatic factishes" signals a date in the history of physics. Henri Poincaré could still distinguish between two different approaches: mathematical physics and experimental physics. Along with them a new type of physics was introduced: theoretical physics. But we are not quite there yet, for we first need to ask how the mobilizing power of the "enigma" came about. What controversies were brought about by the intrusion of the new types of obligations that marked the birth of theoretical physics?

It is very rare for someone talking about science to succeed in inventing questions that scientists themselves have not already asked. The question of the "legitimacy" of the problems facing their colleagues was addressed by Poincaré, as well as by Ernst Mach and Pierre Duhem, physicists for whom the "value" of science was primarily dependent on the sharpness of the distinction they were obligated to make between it and dreams of a knowledge that would reflect the truth of the world. While Max Planck would celebrate, with the conservation of energy, the conquest of a knowledge that transcended human interests, history, and techniques—even "Martians" would eventually produce such knowledge—Poincaré was determined to reduce its scope. "There is something that remains constant," he wrote, in defining energy that is conserved, and no doubt Engels would have accepted this formulation.[2] Even more so as Poincaré was careful to acknowledge the compatibility of his statement with experimental procedure. For him, this "generalization," far from being proven by facts, holds only as a result of its remarkable fecundity: we are allowed to speak of the "principle" of the conservation of energy because, until now, conservation has been a reliable guide for experimentation. But if the requirement of equilibrated energy balances one day ceased to be useful, if it stopped leading to the successful prediction of new phenomena, the physicist would have to abandon this principle, which, although not disproven by experiment, the experiment would nevertheless condemn.

Poincaré was not subject to the hold of enigmatic factishes, nor did he call for another kind of physics. He limited himself to "maintaining the church (of theory) in the center of the village," to struggling so that this church would maintain a living connection with experimental facts. The same was not the case for Mach and Pierre Duhem, who, each in his own way, tried to challenge the arrangement of the village around the "church," that is, tried to abrogate the power of theories, like mechanics, that appeared about to lay claim to a truth that went beyond experimentation. Ernst Mach attacked theories of physics (and to do so he wrote spirited and brilliant historical analyses of mechanics and the science of heat) in order to show that so-called theories are simply an economical summary of forms of practical knowledge, but add nothing to them. Mach appealed to a science that would actively eliminate all reference to the unobservable, whether it involved atoms, absolute space, or, more generally, any terms that would create the illusion that we know how the world is made. As for Duhem, he emphasized that the theoretical structure that organizes the "disorderly mob," the "crowd" of experimental laws, has little to do with logic and, thus, with proof. "Logic provides an almost absolute freedom to the physicist who wants to choose a hypothesis; but the absence of any guide or rule shouldn't disturb him for, in fact, the physicist doesn't choose the hypothesis on which he builds a theory; he doesn't choose it any more than the flower chooses the grain of pollen that will fertilize it. The flower is happy to open its corolla wide to the breeze or to the insect that bears the fruit's generative dust; similarly, the physicist limits himself to opening his thought, through attention and meditation, to the idea that must take root in him, without him."[3] And for Duhem, it is this aesthetics of physical–mathematical creation that must be recognized as the vector of obligations. Duhem, like Mach, appeals to a different physics, a physics that would promote the values of creation—coherence, beauty, simplicity—before those of realism.

Unlike Mach, who adopts a deliberately anti-factishistic strategy, denying that the law can transcend practical facts in any respect, Duhem—and this is the reason for his deep originality, the thing that differentiates his from a classical "antifetishist" position—accepts the bite of the enigma. What he rejects is its localization, the contrast between the enigma of the increase of entropy, for example, and reliable knowledge of mechanics. In other words, what he rejects is that the enigma provides the power and the right to formulate questions intended to dissipate it. For Duhem, all theoretical factishes, whether the product of mechanics or thermodynamics, are equally enigmatic. That is why the "abstract" character of thermodynamic potential, which defines a state without actually describing it, and the Carnot–Clausius transition from one equilibrium state to another, which artificially subjects irreversible processes to a rational norm that is foreign to them, far from inspiring any frustration, are fully satisfying to him.

Duhem also appeals to another physicist. This physicist-mathematician would know that, along with the entire physical-mathematical edifice, none of the propositions that constitute it can be confronted with experience. Faced with an experimental contradiction affecting some consequences of a theory, she would be aware of the fact that she can choose to modify the edifice or destroy its foundations. She would be particularly aware of the fact that, whatever her choice, although no one could say she was wrong, no one could say she was right either. Such a physicist could never refer to the satisfaction of experimental requirements to abdicate a responsibility that no logical method could help her satisfy. She would work within the grip of an enigma but would realize that it was not up to her to see through it, to ensure, or even hope to ensure, that reality confirm the convergence between its "reasons" and those presented by theory.

There are very interesting differences between Poincaré, Mach, and Duhem, those witnesses of the moment of "hesitation," when, a century ago, theoreticians questioned the strange beings they had brought into existence. These differences mark them as true authors, negotiating the meaning of the commitment to which those beings obligate them. But these authors have one thing in common: in no case can they be confused with the conventional epistemologists, positivists, or instrumentalists who would subsequently claim them as their own. In no case do they address some "science in general," which could be defined according to the model of physics they constructed and which would thus be ensured of its rationality.[4] All of them attempt to interpret the obligations brought on by the new situation of physics through the creation of beings that seem endowed with the power to impose their own questions, to judge phenomena in terms of requirements that transcend experimental evidence. And in doing so, all three of them reject what might be called the "grand positivism" of the period, that great imaginative vision inhabited by the trust and hopes brought about by a twofold unification: the unification of physical–chemical nature by energy and the unification of "historical" nature by a more or less Darwinian evolution.

Like Poincaré, Mach, and Duhem, the "great positivists" such as Wilhelm Ostwald, Herbert Spencer, Ernest Solvay, and Ernst Haeckel were not the victims of the history of physics. They were defeated fighters, the vanquished protagonists of that history. You cannot be a victim when you're the author of a thesis claiming general validity. But their respective defeats are very different. The dream of Ostwald, Spencer, Solvay, and Haeckel did not die with them. Quite the contrary, for it comes back to life whenever a new concept appears to promise a unitary conception, where the intelligibility of nature produced by the sciences and that of the biological-social evolution that is ultimately productive of those sciences seem about to come into contact. We need only consider the contemporary appeal of "complexity" theory. On the other hand, the most eloquent symbol of the defeat of Poincaré, Mach, and Duhem, the reason for my interest in them, is that their work is now considered to be a part of epistemology, read by philosophers of science and not by the physicists to whom it was addressed. Correlatively, the "lucidity" they defended—each in his own way—against the fascinating enigma has become a questionable virtue for physicists, synonymous with "defeatism," treason, and the cowardly or positivist fear of going beyond phenomena, or the questionable desire of limiting oneself to "saving" them in some coherent fashion.

Psychology and epistemology don't have the ability to explain how a scientist decodes her obligations. When analyzing scientific conflicts, it is always in retrospect that the power of psychological or epistemological categories becomes relevant, for this apparent power signals the fact that the problem presented to the actors and by the actors no longer interests anyone except the epistemologist, psychologist, or philosopher. That the history of science "in the making" is not subject to these categories doesn't, however, prevent a given actor from assigning them a certain relevance. But if, during a controversy, a scientist adopts a long-term historical or epistemological perspective, her position should not be analyzed in terms of that perspective. On the contrary, we need to address her position in order to understand how that perspective was able to become an argument for her. For example, Pierre Duhem discussed the considerable hesitation that has marked physics, ever since Copernicus, between "saving" phenomena, that is, reproducing them in a mathematically coherent fashion, and "explaining" them. But for Duhem, this is exactly what was at play in the difference between thermodynamic and mechanical potentials.

Whenever such historical comparisons arise, together with a reliancc on the key topics of the epistemology or psychology of knowledge, they most often translate and betray a situation wherein one of the ingredients can be viewed as a conflict of obligations. That is why their relevance is always local and circumstantial: it's possible that the same scientist might, without the slightest hesitation, move between one perspective and another; for example, from a discourse focused on the lucidity and limitations that theoretical constructions must obey, to one that makes prominent the most triumphant realism, and vice versa. These all-purpose general "values," which appear to explain or justify her position, entail no obligations; they merely generalize what the theoretical being she was in the process of constructing obligated her or committed her to.

The "epistemological" crisis in physics a century ago does not, therefore, reflect the past, when "physicists were still philosophers," it is contingent upon the singular history of physics. What would have happened without this strange offshoot of the caloric theory and rational mechanics, the Carnot cycle? No doubt the two "laws of thermodynamics" would not have been formulated and would not have brought about, through their obvious parallelism, the ambition to confer upon the "degradation" of energy a scope comparable to that of its conservative transformation, nor, for that matter, the rejection of that ambition.

But how and why did the confrontation among physicists cloak the meaning we have inherited? For the challenges that divided "realists" and "rationalists" among physicists have been more or less forgotten, whereas the present moral of the story, the one the operation of anamnesis I am attempting seeks to resist, now leads us to "expect" a solution to the enigma, the triumph of the Queen of Heaven, the triumph of a unified conception of the world according to which the second law of thermodynamics will become simply a question of probability.

Needless to say, what I have just described is a necessary ingredient in this matter, for it would make no sense if irreversibility hadn't first become the key property, capturing the multiplicity of physical–chemical processes and the problems they are liable to introduce. But how, then, did the "arena" created by the Carnot cycle and the protagonists who defined themselves according to its terms flip into the category of what is now referred to as "classical physics," crisscrossed by merely epistemological conflicts between "positivists" and "realists?" How was the page turned so that all their questions appeared to be determined by what they ignored: the major challenges of twentieth-century physics, the revolutions that enabled physicists to tell us that "humankind" is finally faced with the question of what it can know?

It is when we arrive at questions of this kind that anamnesis has done its work. For it will have re-created a problem where an impression of progress was once dominant, and blended together what had once been dissociated into epistemological analysis and strictly physical production. It will also have managed to make interesting, because surprising, the requirement that was taken up by physicists throughout the twentieth century: we require that our faith in the realist value of physics be recognized and respected, and we are satisfied with our theories only if they confirm that faith, only if they do not lead us to betray the vocation of the physicist.

What is interesting and surprising is that the scene I introduced to focus the question of this vocation in Book I, "The Science Wars," which pit Max Planck against Ernst Mach, has now lost its character as a model, becoming something that should not be repeated. Now, both Planck and Mach are situated in a precise historical moment. The "faith" of the one and the logical–historical criticism of the other are no longer "grand alternatives" that transcend history, but the polemical expression of a history that, in 1908, was in the process of ending in a singular manner, that of a general mobilization around an alternative that forced everyone to take sides. Similarly, the opposition put forth a few years later by Einstein, the opposition between the desire to escape the vicissitudes of the world that leads the "true" physicist to the temple of science and the utilitarian concerns that those who will only be parasites follow,[5] superimposes on a completed history a morality that seems to have endured throughout the ages but primarily reinterprets the closure and disregard of the crisis. For it is indeed a crisis of physical reality that has come to pass, in the sense that its "moralization" refers it to a dead past. Ostwald, Duhem, and Mach, each in their own way the vanquished in this history, were "revolutionaries," struggling—somewhat like the future creators of quantum mechanics, Bohr, Heisenberg, and Pauli—for a new conception of the history of physics, one that would extract itself from the particularity of its first objects. They did not object to "realism" out of utilitarian conviction. They stated that it was the deceitful simplicity of its first, mechanical objects that promoted the "realist" physicist's naive belief in a reality generally capable of dictating its own reasons. And they saw thermodynamics as a generalization of dynamics, forcing it to accept the lucidity it had until then by and large avoided, obligating physicists to say good-bye to the transparent and rational world where the Queen of Heaven reigned supreme.

The retroactive history of the sciences is frequently unfair to the vanquished, but the way in which it is unfair is highly significant. In this case, the fact that an "antiutilitarian" morality could be deduced from the episode once it was over is not simply a case of the victor kicking the vanquished when he's down. It enacts a question that, following different modalities often depending on the country, certainly electrified the crisis of physics and ensnared it in a network of political and cultural imbroglios.

In this respect, the history of Henri Poincaré is significant. Poincaré had written *Science and Hypothesis* in 1902, with little foreboding as to its repercussions. But to his great surprise, his claims were mobilized for the controversy on the "bankruptcy of science" that had been raging in France since 1895. Against the "Third Republic" alliance forged between the values of science, the laity, and the Republic, for which Marcellin Berthelot was a perfect illustration, the Catholics claimed that, as a source of moral values, science was bankrupt. And how could it not be, they added in 1902, since, as Poincaré had shown, it is finally nothing more than a collection of "convenient recipes" aimed at action and prediction rather than veridical understanding. The first lines of Poincaré's next book, *The Value of Science* (1907), reflect the author's indignation: "The search for truth should be the goal of our activities; it is the sole end worthy of them." And the entire book can be read as a protest against the distortion of which he was a victim. But the fact is that the debates among physicists concerning the status of physical theories involved other protagonists, and these had little concern for subtlety. They put the physicist's back against the wall: did he or did he not believe that the laws of physics told the "truth"?

Although the peaceful Poincaré may have been pulled into the fray, how could the polemic-loving Duhem avoid it? In fact, Duhem did not even need to take a position. His claims were recognized by the "laity" as being dangerous, and he was the object of a forceful attack by the philosopher Abel Rey, who accused him of sapping the confidence of the population of silent, hardworking physicists.[6] To the extent that a mechanical reality is a knowable reality and allows us to state that "empirical reality" directly authorizes theory, it inspires confidence and faith, and the physicist at work is therefore "spontaneously mechanistic." Doesn't the skepticism and mathematical subtlety of a man like Duhem reflect the pernicious desire to destroy the nonreflexive confidence that is the physicist's strength?

For Rey, the mechanical conception of reality was merely a *practical postulate* that privileged models that made use of bodies in motion, and not a truth in the philosophical sense. That is why Lenin, in *Materialism and Empirio-Criticism,* disqualifies him as a "confusionist," an unacceptable attitude in the crusade he had just begun against the "Machians" Adler and Bogdanov: the future of the revolutionary movement appeared to be connected with the values of knowledge.

While the positions of physicists concerning their theories were used in arguments in which the values of civilization and human history were in question, the question of their freedom of action, their autonomy, and their means—their working conditions—also led to a confrontation between the values of knowledge and the values of "civilization." During the second half of the nineteenth century, the requirements that scientists could impose upon the government (to finance research) as well as the requirements that government and industry could impose upon science (to focus on questions that had economic importance) had become part and parcel of offensive and defensive strategies. In 1863, the German chemist Liebig, under the pretext of writing a book on Lord Bacon, engaged in a virulent attack on an English science dominated by utilitarian values, and presented a plea for the autonomy of a science preoccupied solely with truth—and that would be truly useful to society if and only if it were not constrained by short-term considerations.[7] During that same period, French scientists were fascinated by the symbiosis between science, government, and industry that seemed to characterize Germany, but the symbiosis troubled them as well. For didn't French science, as poor as it was, benefit from an inventiveness that the Germans, subject to utilitarian interests, were forced to renounce?[8] In short, this gave birth to what would become a constant preoccupation in the twentieth century: the defense of the autonomy of science in the face of economic, industrial, and government interests. The role played in this defense, ever since Planck and Einstein, by the "psychosocial type" of the inspired physicist, productive if, and only if, she is free to pursue her vocation, is well known, and its message is always the same: a disinterested science is the "goose that laid the golden egg" for civilization, a civilization that must therefore avoid asking it for more than it is capable of giving. This sorry metaphor points to an academic institution that abandons "applied" research and its subjugation to socioeconomic requirements. It marks, along with the figure of the inspired physicist, a moment when the values of "progress" that refer respectively to civilization and the sciences have ceased to coincide. The social and economic importance of scientific research is now a threat to the vocation of scientists. Regardless of the twists and turns of history, we should not forget one thing: the triumph of physical laws and of Maxwell's demon also signified, for physics, the end of any compromise with "utilitarian values." For physicists, to go "beyond phenomena" also meant escaping the threat of being seen as simple instruments of technical and economic progress.

There was thus at work, at the end of the nineteenth century, a series of "weighty causalities" that together produced what can be called a "critical milieu." Here, the term "critical" is understood in the only interesting sense capable of connecting the crises of human societies with the "critical states" described by physics: the disappearance of distinctions of scale, the multiple resonance of normally separate dimensions. And a particular question allows the series to converge and cause the dimensions to resonate, not in the sense that it could lead to a solution, but that it serves to promote a cause that would transcend any crisis. Although to a large extent the problems are shared by many of the other sciences, chemistry in particular, physicists alone are in a position to present themselves as defenders, not of the generic cause of disinterested science, but of the particular one of an inspired physics, the only kind capable of unraveling the enigma of reality.

In Book I, I put forth the idea that the way in which a science is presented, the way in which it defines its connections with other practices and with reality, is part of the identity of that science. The fact that physicists have sought a kind of "ostentatious" display of the cause they defend, whatever the fate of the other sciences, constitutes a change of identity. To this change corresponds the exceptional role that has been assigned to their science: it is physics that, in various ways, has been seen as a witness or a target; it is its relationship to reality and truth that has been repeatedly commented on, as if it were the crucial site around which the always recurring question "what can we know?" was to be debated. On the other hand, the theme of the "vocation of the physicist," by the brutality with which it breaks its ties with other so-called rational practices, reflects the inability of rational argumentation to defend the hopes and claims associated with an inspired physics. The time for the subtle prudence of Poincaré or the ascetic lucidity of Duhem was long past. It was time for slogans and choosing sides.

The above contextualization can certainly help clarify the virulence of the operation of mobilization through which the hypothesis of a "rational" physics that cultivates the virtues of the most ascetic lucidity has been bracketed, much like a parenthesis. But it does not allow us to understand how this parenthesis has been closed. It only allows us to anticipate a link between the new identity assumed by the physicist and the way in which the identity of physics will be forged, what it will privilege, what it will treat as secondary. But it does not enable us to deduce how that link was created. Once again caution is required so that we not turn the "vocation of the physicist" into a simple image that will then be referred to as "ideological." Physicists would have reason to complain and argue that *something happened to change the stakes.* This explains why Planck, for example, former fervent defender of the distinction between dynamic and thermodynamic states of equilibrium, became a proponent of a unified vision of the world. Something happened that, again, singularizes the history of physics and can help explain why, for any contemporary physicist, the history of physics could not under any circumstances agree with the vanquished. By who or what were Duhem, Mach, and Ostwald defeated? Atoms did it, any physicist would answer.

19

Percolation

Atoms! It was nature in person that tipped the balance, and in doing so it assumed the oldest speculative form with which modern physics recognized some kind of connection. And what is strangest of all is that this response, which is true as a first approximation, when history is seen in terms of progress, *remains true* in successive approximations. Providing, that is, we inquire into the relation atoms might have to the triumph of the Queen of Heaven. For, in the guise of an old actor no longer in fashion, seen by rationalists like Ernst Mach as a mystifying representative of the belief in a world knowable "in itself," it was indeed a new protagonist that appeared onstage, associated with the new practices that brought it into existence. And nothing is less obvious than the way the relationship between these new practices, these new actors, and earlier history was constructed.

Something "had" to happen at the end of the nineteenth century, independently of the "weighty causalities," cultural, political, and economic, that I discussed above. For, experimental physics was now in the process of creating the means to access unobservable actors, beyond phenomena—ions, electrons, nuclei—that in one way or another "had" to transform the identity of physics. "In one way or another"—this point is important. For physicists of the new generation—Paul Langevin, Jean Perrin, Albert Einstein—the regular, "macroscopic," phenomena that were the shared heritage of mechanics and thermodynamics were now part of an outmoded past. The triumph of mechanics, Maxwell's Queen of Heaven, was far from being reassured by the existence of "atoms." Why, then, did the turning page of history not push aside the Queen of Heaven and the demon, the heirs of Hamilton and Maxwell, the way it swept aside the descendants of Carnot and Clausius?

We shouldn't expect an answer that, like a magic wand, suddenly resolves the problem and makes any confusion disappear. We are dealing with history, not with logic, and perhaps the best model for our purposes might be that of "percolation," proposed by Michel Serres. There is no single reason, no cause that in itself has the ability to assemble and lead, no shepherd to gather up multiple causalities, for by themselves they will never get us very far. But connections between them may become locally denser until, at some point, a threshold is reached and things begin to "flow": "In fact, like the Amur or the Yukon rivers, history filters, abandons, retains, returns, forgets, lazes around, freezes, or seems to sleep among multiple traceries, and suddenly, without our being able to foresee it, brings about a linear flow, a straight line, irresistible, nearly permanent, as if immortal."[1] History appears to "flow from a source," making its way through the undulatory mechanics that Louis de Broglie associated with Hamilton, then through quantum mechanics, which confers upon a generalized Hamiltonian formulation the ability to represent stationary states of the atom.

There is a richness to the percolation model because it is able to recognize the interplay of general causes (gravitation, whether for a coffee pot or a river) without allowing them to eliminate narrative or transform it into a moral tale. History flows, but other historical possibilities "dry up," incapable of resisting the grand narrative taking shape, and it is all those absent histories, all the questions that weren't asked or were left unanswered, that delineate the true space of percolation. Here, although we speak of "weighty" causes, referring to the "socio-cultural context" or "experimental physics," we simply mean that "something had to pass, something had to happen." But the truth that will become permanent, almost immortal, once the physicist's vocation has become stabilized, the truth that illustrates and celebrates the triumph of the Queen of Heaven over thermodynamics, will not have the power, after percolation, to resolve the questions that were the critical bread and butter of the previous generation. These will have "dried up." But the questions will return again with a new set of circumstances that will fascinate new generations of physicists. Does Schrödinger's cat die in its sealed box when there is no one around to observe it?

Before continuing that history,[2] I want to track the way the course of history thickened around a modest strand. For the hypothesis claiming that heat was nothing other than an invisible form of motion of the constituents of matter was ancient, and hardly prestigious. Associated with Bacon, Boyle, Hooke, Leibniz, then Rumford and Davy, it produced no practice of measurement, unlike caloric theory, and it was derided by calorists as sterile speculation. Yet, it was always available, and when the conservation of energy killed the caloric theory, James Joule, von Helmholtz, and others immediately referred to it as a promising alternative to the theory of heat-as-substance. Yet, it was Clausius, in his article "On the Type of Motion We Call Heat" (1857), who was the innovator. Once again, "collisions," which had been overlooked ever since rational mechanics had made them subservient to dynamic conservation, were introduced. However, what was coming back was no longer "the" collision event that had been associated with the question of a possible loss of motion. Collisions were now elastic, and therefore conservative, and they played roles as anonymous parts of a multitude. And it was this multitude of collisions that was called on to explain what a gaseous state of equilibrium "is," and how that equilibrium is (irreversibly) achieved. The "kinetic theory," as it was called, is indeed an event in the history of physics, for it enabled a redistribution of the relationship between "state" and "explanation." The state (of thermodynamic equilibrium) was no longer self-explanatory, it was the result of an innumerable and tumultuous reality that explained it but that, at the same time, made way for the theorization of the familiar increase of entropy toward equilibrium about which thermodynamics is silent.

The discussion first turned toward the kinetic model, which was not a theory. Clausius's analysis (and Kronig's, who published just before him) primarily involved gases and was limited—an intentionally partial representation—to assuming that molecules underwent translation interspersed with elastic collisions. Clausius recognized the likely existence of other motions affecting molecules, but translation was sufficient to interpret the expansion of the gas and all the relationships between pressure, volume, and temperature ($pV = nRT$) that were formerly part of the physics of caloric. Those relationships, as mentioned earlier, characterize states of equilibrium. Their nature is "purely experimental." This allowed them to survive from the time of Boyle's earliest measurements, to acquire remarkable precision within the context of caloric theory and resist the discovery of the conservation of energy.

Clausius used the phenomenon of evaporation in a sealed enclosure, characterized by a known equilibrium between two "phases": some of the liquid remains liquid and some enters the gas phase. In molecular terms, entering the gas phase means that the strong mutual attraction among molecules in the liquid does not prevent some molecules from escaping. These liberated molecules now move freely and are affected only by collisions with other gas molecules. But some gas molecules may also be trapped by the attraction of molecules in the liquid. Equilibrium is established when, on average, as many molecules are trapped per unit of time as are liberated into the gas phase.

What is essentially novel about kinetics can be found in this description and the creation—from Clausius forward—of a new equivalence that now results from *statistical* compensation. It is important to emphasize the difference between this equivalence and the one common to dynamics and thermodynamics. The latter creates an identifying equality, independent of which the terms it unites are stripped of physical meaning (why mv^2, why Q/T?). Statistical compensation, however, implies an elementary activity that is left undefined. Collisions occur continuously in a gas and are not, in themselves, different whether or not the gas is at equilibrium. The state of equilibrium is nothing more than a simple outcome of this elementary activity. The true subject of the description is now a disorderly multitude. The Brownian motion of a particle of dust, whose direction varies wildly from one moment to the next while the particle is suspended in a liquid apparently at rest, will be chosen by partisans of the kinetic hypothesis as the visible witness of this new relationship between appearance and reality. The liquid appears to be at rest, but the Brownian particle, subject to countless shocks, *reveals* the incessant agitation of constituent molecules in that liquid. What nature forces us to envision is agitation where we perceive rest. But we are not obligated to follow the motion of every individual particle; what matters is the average effect and, therefore, the relative *frequency*, of the different types of events that contribute to what we observe.

Clausius's kinetic model affirms its dependence on the science of motion but, at the same time, ensures that dynamic equivalence is not relevant: forces, acceleration, potential energy—everything that can be used to identify the state of a dynamic "system"—are not denied, but they may be ignored.

The use of statistical compensation only takes into account discrete events, which can be defined as random because only the average result of their behavior is taken into consideration. Correlatively, the relationships of reciprocal determination between variables that characterize the thermodynamic equilibrium state become intelligible. How does a change of temperature result in a "shift" of equilibrium between liquid and gas phases? If what temperature measures is the average speed of molecules, or the average energy that characterizes their motion, this is understandable. The expulsion of a molecule outside a liquid is an energetically "costly" event because the molecule takes with it, in the form of motion, part of the energy associated with interactions among molecules in the liquid, and the probability of this event depends on the degree of agitation of those molecules, that is to say, the temperature of the liquid. But the probability of trapping a gas molecule and restoring its energy to the liquid depends on the number of gas molecules. The higher the temperature, the greater the number of molecules capable of escape, and the greater their number, the greater the frequency of their capture. Therefore, the increase in temperature intensifies the two rival processes, and it is this complex relation that the thermodynamic laws of equilibrium translate.

The most celebrated convert to the kinetic hypothesis was undoubtedly James Clerk Maxwell. The history of his conversion reflects the singular fecundity of the symbiosis this hypothesis was able to bring about between the exploration of new experimental, that is, macroscopic relationships, and the construction of precise properties attributable to particles.

In 1860, Maxwell went beyond the hypothetical approximation adopted by Clausius, which assigned the same average velocity to all the molecules in a gas, and centered his description on the concept of *statistical distribution,* which characterized an instantaneous state. Each state was defined by a known distribution of velocities among the molecules in a gas, and the gas at thermal equilibrium corresponded to the situation where the incessant collisions among molecules, which continuously modified their velocities, no longer changed the distribution of velocities. In other words, microscopic activity, rather than being assigned an average velocity deduced from macroscopic observables such as temperature, was presented as productive of this average. The description of equilibrium effectively made use of statistical compensation: at every instant, approximately as many collisions occurred causing a molecule with a given velocity to "disappear" as collisions that caused a molecule with that same velocity to "appear," and this held true for all velocities. Maxwell showed that, according to his model, the viscosity of a gaseous fluid should vary with temperature but not with the density of the gas. In 1866, the experimental proof of this conclusion, which had initially left him skeptical, transformed Maxwell into a believer of what he significantly referred to as the "dynamic theory of gases."

Recall that Maxwell jeered at the "German Icari" like von Helmholtz, Clausius, and Boltzmann, who sought to identify the irreversible transition to an extreme of thermodynamic potential with a law of Hamiltonian dynamics.[3] He was familiar with the "statistical considerations" used to link macroscopic description and molecular interaction and appreciated their value, but they were only good for interpreting those cloudy forms, the billions and billions of molecules whose dynamic motion there was no hope of following. There, where the Queen of Heaven explicitly affirmed her reign, where Hamiltonian dynamics provided a complete theory, statistical arguments had to disappear, and with them the irreversible growth of entropy. For Maxwell, after 1870, kinetics was nothing other than an incomplete theory of dynamics.

Is kinetic theory, then, merely the humble servant of dynamics, a valid approximation certainly, but one that refers to a world of individual dynamic trajectories that a demon would be capable of following? This is the conclusion to which we have become accustomed, and it is the conclusion heralded by Maxwell's demon. According to Thomson, this demon, because it knew how to select, conserve, and use the energy of molecules in motion, could force a chemical reaction to occur. He writes: "Let him take in a small store of energy by resisting the mutual approach of two compound molecules, letting them press as it were on his two hands and store up energy as in a bent spring; then let him apply his two hands between the oxygen and double hydrogen constituents of a compound molecule of vapour of water, and tear them asunder."[4] And that's that. The reaction takes place, and in doing so has manifested its submission to dynamic intelligibility. But—and this is the point I want to make—Thomson was not seeking, in chemistry or elsewhere, the means to prove that this submission was effective. He assumed this to be the case. What he wanted to show was that the second law of thermodynamics, the macroscopic law of "irreversible degradation of energy," did not hold true for the demon. It could cause chemical reactions at will and, if it so desired, cause a system to deviate from a state of chemical equilibrium.

Maxwell and Thomson, therefore, already "maintain" the statistical interpretation of entropy but don't allow us to understand the process of percolation itself, the fact that this interpretation was accepted as settling, in its generality, the question of the relationship between the "dissipative" world of physical-chemical processes, for which the kinetic theory is relevant, and the conservative dynamic world, with respect to which kinetics is presented as an "incomplete dynamic theory." In fact, one question is not asked that should have been. Let's forget for a moment the kinetics of gases, or changes of phase, for in their case dynamic interactions (attraction and repulsion) appear capable of giving meaning to kinetic events, that is, collisions. What, then, of the disparate crowd of kinetic events required to understand the equally disparate crowd of "energetic" processes?

Chemical reactions were very quickly interpreted (in spite of the objections of "true" thermodynamicists such as Pierre Duhem)[5] in terms of "reactive collisions" among molecules, collisions that did not result in a simple change of velocity but in a chemical transformation of the molecules. The state of chemical equilibrium would then correspond to a statistical compensation between the effects of different types of reactive collisions, some creating a compound, others destroying it and restoring the original reagents. But what is a "reactive collision" in dynamic terms? Naturally, there was not even the hint of an answer to this question. In the early twentieth century, it was the turn of radioactive disintegration to be subjected to kinetic calculation by Ernest Rutherford. Radiation at equilibrium was related to the respective lifetime of each kind of radioelement in the decay chain of the initial radioactive product. But what is a radioactive decay event in dynamic terms? No answer. And in 1916, it was to kinetics that Einstein turned for the secret of the absorption and emission of light by the Bohr atom. He showed that two types of events are necessary: those induced by an electromagnetic field and "spontaneous" events that are characterized, like radioactive decay, by a lifetime. Why did Maxwell's demon survive when no theory could show what he was supposed to do to "undo" a process characterized by a lifetime?[6]

Another history of physics might have been possible, one that seems to have been foreseen in Jean Perrin's *Atoms*. This history would be centered on the event and would have sought out the secret of the diversity of physical–chemical transformations, not postulated their unity based on the fact that they all conserve energy. And perhaps, in this history, the discovery of the existence of spontaneous events would have been narrated as the moment when physics had to finally say good-bye to the ideal of intelligibility associated with the Queen of Heaven. In fact, the very notion of a spontaneous event would seem to destroy the reversible equality between cause and effect. The disintegrated nucleus would never be reconstituted; a Bohr atom, once spontaneously "de-excited," would never be spontaneously re-excited without induction by an electromagnetic field. In such a history, Duhem's thermodynamics would have been vanquished, as it was in ours, but it would have been vanquished by kinetic events, not by Maxwell's demon. And that strange statement would never have been heard, wherein physics seems to claim the ability to deny what it is describing, a statement that, at this stage of our story—and this point needs to be emphasized—we still don't have the means to understand: from a "fundamental" point of view, the temporal asymmetry of processes does not exist, it is merely a simple matter of probability.

By describing the percolation process, we introduce the problem of the justifications given by a history that has "flowed" past. The kinetic interpretation of Clausius and Maxwell appears to justify the demon, but the argument doesn't get us very far. One plausible possibility that would have confirmed the diversity of events has not been actualized. Historically, kinetics has not been defined as the science of events but, as Maxwell proposed, it has become a subordinate science, an "incomplete dynamic theory." Events were not recognized as introducing a new problem, requiring a new intelligibility and giving rise to new obligations on the part of physicists.[7] Events were associated with the model of elastic collision, which is ruled by the theory of dynamics but reduced by kinetic theory to a black box that transformed molecules entering it with given velocities into molecules leaving it with different velocities. Similarly, every kinetic event was seen as a black box whose secret the theory of dynamics would one day reveal.

We now arrive at the conclusion of anamnesis, an episode from the textbook history of science: the construction of Boltzmann's well-known $\mathcal{H}$ theorem. We know that the subsequent history of this development will confirm Maxwell's verdict without enabling us to "justify" the percolative flow that claims to follow from this confirmation. Boltzmann's theorem accepts the identification between the kinetic event and the dynamic event, and will contribute nothing to clarifying the reduction of one to the other. But this was the theorem that entered history and it is that history that brought the twentieth-century physicist into existence.

What was Boltzmann looking for? We know he had already attempted to turn the second law of thermodynamics into a mechanical theorem. But we would be wrong to think that, in doing so, he was trying to confirm the reductive ability of mechanics. What interested Boltzmann, like Maxwell before him, was the relationship between macroscopic and microscopic properties, and the possibility of creating new kinds of models that would extend the field of theory to experimental relations that had not yet been interpreted, or were even unknown. In this instance, Boltzmann wanted to extend Maxwell's work to investigate an area about which thermodynamic theory was silent: the field of nonequilibrium processes, especially relaxation phenomena (which refers to the return to equilibrium of a disturbed medium). That is why there is nothing unusual in Boltzmann's addressing the simplest case, that of a "dilute gas" (low density), where "events" present no particular problem because they are represented as elastic collisions. Only by using an example such as this could he hope to demonstrate the fecundity of kinetics where a purely macroscopic theory had failed, and possibly succeed at last in constructing the representation of a process with increasing entropy. To succeed, in other words, in getting the enigmatic factish represented by increasing entropy to "talk."

Boltzmann's theory derives from an equation that introduces the change over time of the statistical distribution of the velocities of gas molecules. Maxwell had defined the equilibrium value of this distribution. Boltzmann wanted to describe the change leading to equilibrium. The equation constructed by Boltzmann is an "integro-differential" equation, the physicist's nightmare, the translation of the complicated relationship between state and change corresponding to the kinetic hypothesis: the variation in the distribution of velocities at a given moment is the result of the integration over all velocities of the effects of the different collisions at that moment, but the relative frequency of those different collisions is itself a function of the distribution of velocities. However, when the appropriate approximations are made, the equation yields valuable information, especially about transport properties (viscosity and heat conductivity) and relaxation time. In that sense, it was the first in a line of kinetic equations that remain an integral part of statistical mechanics to this day. Moreover, Boltzmann's kinetic equation paved the way for a general conclusion, which turned out to be extremely valuable to him. It allowed him to construct the well-known $\mathcal{H}$ function, which had the property of decreasing to a minimum whenever equilibrium was achieved. For Boltzmann, this was a moment of triumph. For, while the collision between two particles is represented as a strictly dynamic event, collisions within a population of particles can explain the irreversible approach toward equilibrium.

The story that follows has been told a thousand times. Although Boltzmann presented his equation as being strictly derived from mechanical description, he was later forced to recognize that he had introduced a hypothesis. This hypothesis was essential for his calculations. But if it was identified with an approximation, then the irreversible increase of entropy—or the decrease of $\mathcal{H}$—predicted by his equation was condemned to remain "merely" a result of our ignorance rather than being attributable to nature. This hypothesis is that of "molecular chaos." It came into play when Boltzmann determined the average number of different types of collision from the distribution of velocities at a given moment. Because that distribution concerned only velocities and ignored the relative positions of the particles, the validity of the operation depended on the fact that those positions didn't matter, that is, on the fact that knowing them would not provide the observer with any additional insight. In technical terms, we say that in order for the hypothesis of molecular chaos to be valid, particles must not be "correlated" before the collisions. Acknowledgment of the limitations of Boltzmann's model is only retroactive, however. It dates back to the last decade of the nineteenth century, at a time when Boltzmann had been defeated and when it became a question of knowing what had defeated him.[8] In this case, it meant understanding why his theorem, apparently purely mechanical in nature, was vulnerable to the objection raised in 1876 by his colleague Loschmidt.

What is traditionally called "Loschmidt's paradox" introduces a very hypothetical and perfectly unrealizable "instantaneous reversal of the velocities of all the particles" in a Boltzmann gas. In this case, according to the dynamic equations, each particle supposedly retraces its path in reverse and the "reverse" collisions "undo" what the "direct" collisions accomplished. Consequently, if the state affected by the reversal characterized a gas moving toward equilibrium, the state resulting from the reversal would be part of an evolution that shifted the system away from equilibrium, and the corresponding magnitude $\mathcal{H}$ should increase rather than decrease. Boltzmann's theorem thus provides, quite inadvertently, a recipe for obtaining an "antithermodynamic" change, one that is prohibited by the second law of thermodynamics. To contradict the law, "all" that is required is to begin from an initial state obtained by reversing all the velocities of a given state of a system approaching equilibrium.

Loschmidt's paradox revealed the hidden weakness in the construction of Boltzmann's $\mathcal{H}$ function, that is, the hypothesis of molecular chaos. The unique feature of the "reverse" collisions that follow from the reversal of velocities is precisely that they do not satisfy the hypothesis: the ability of these collisions to bring about a decrease in entropy is strictly dependent on each molecule being characterized not only by a precise velocity, but also by a precise position relative to the other molecules. If the positions are slightly disturbed, the gas will "normally" move toward equilibrium. In other words, after the reversal, the molecules must be described as correlated. Not only can the complete dynamic theory predict these correlations, which would always be the case, but in their case the theory is necessary for predicting the "abnormal" change the collisions among correlated molecules bring about.

This would be an appropriate moment to stop and assess the situation, for it is quite surprising. Why didn't Boltzmann object that Loschmidt's argument was not an acceptable argument in physical terms? Who was going to perform the reversal of velocities and show that Boltzmann was wrong? No manipulative demon was up to the task, for it would have to have had 10^{23} hands to simultaneously manipulate *all* the constituents of an instantaneous state. It was apparently the Queen of Heaven herself who showed up to promote the most incommunicable of her attributes: the symmetry over time of the equations of dynamics. But wasn't there something nearly supernatural about her authority? Didn't she supply the theoretical equations that all physicists recognize, at least formally, as deriving their legitimacy from experiment, with the ability to assert, contrary to experience, that it was entirely conceivable for a system to spontaneously move away from equilibrium? Boltzmann couldn't make use of an argument of this kind because he would thereby have acknowledged that his own adversaries were right. It was Ernst Mach who supported the necessity of ensuring that theory fell within the strict framework of the operational limits that qualify physics as a science. Boltzmann pleaded for the freedom of creative imagination, one that could build models, but he was defeated by another use of the imagination—imagination in the service of power rather than creation.

What is also surprising is why it wasn't until 1876 that Loschmidt's objection was formulated. Didn't it already follow from the very distinction between reversible and irreversible processes? For us, certainly. But unless we believe that nineteenth-century physicists were incapable of reasoning like a first-year college student of today, we have to assume that the property of time symmetry that characterized dynamics *had not been explicitly formulated at that time.* And in particular we need to bear in mind that, until then, neither reversibility nor irreversibility had been recognized as reflecting the symmetry or nonsymmetry of physical changes over time, what has since been referred to as the "arrow of time." The contrast between dynamics and thermodynamics was part of the development of rational mechanics, and based on notions of conservation and loss. It was the absence of loss, the full equality of a cause in its effect, and not reversibility over time, that conservation then referred to and that was illustrated by the thought experiment of the Galilean ball returning to its initial height. And it was the question of the degradation of energy, the spontaneous leveling out of differences, that Maxwell's demon challenged. It was only after Loschmidt's objection that a new conclusion became obvious: we cannot relate the irreversible evolution toward equilibrium, the "degradation of energy," to human limitations and ignorance without also relating all our physical observations concerning the difference between past and future to that same ignorance. Dynamics became the science of reversible change, no longer in the conservative sense that relates it to the concerns of engineers, but in the sense that such change is indifferent to the distinction we make between before and after.

The way in which this transformation of meaning occurred, the fact that it was in order to bring down a "German Icarus" in midflight that the most incommunicable of the Queen of Heaven's attributes was finally made explicit, are some of the questions raised by "historical percolation." Who knows what effect this argument would have had if it hadn't been presented as an objection, highlighting the power of the Queen of Heaven to "undo" the evolution toward a state of increased entropy? What if it had been presented as a polemic, challenging the absurdity by trying to extend the jurisdiction of the Queen of Heaven to the field of earthly, dissipative physics? In any event, what has now stepped onto the stage is the one attribute of hers that will retroactively become the key to the relationship between dynamics and thermodynamics. To subject the second to the first, to acknowledge the subordination of all "phenomenological" physical chemistry to the rule of "more fundamental" laws, it is not necessary to explicitly subject each type of kinetic event to those laws. It is not even necessary to know the laws to which they respond. All that is necessary is to state that all kinetic events are, like elastic collisions, reversible over time.

In 1877, Boltzmann accepted the objection based on the reversal of velocities. He had indeed been backed against a wall by the property of symmetry of the dynamic equations that are made explicit by Loschmidt's thought experiment. The possibility of a dynamic state like the one resulting from an inversion of velocities and the fact that such a state would result in a change that caused entropy to decrease seemed to him irrefutable—much to his chagrin. All he managed to say about such a possibility was that it was "highly unlikely." Boltzmann then went on to develop the probabilistic interpretation of entropy that has been associated with him ever since, the corresponding equation even being engraved on his tombstone. Contrary to the $\mathcal{H}$ theorem, this interpretation makes explicit appeal to our ignorance, and correlatively lacks any ability to describe a temporal change. Every dynamic state is, in effect, considered "equally probable," and the probability of a macroscopic state is measured by the number of distinct microscopic states that can be characterized in terms of the corresponding macroscopic description. This was a new definition, a combinatory definition this time, of the state of equilibrium: it is the state resulting from the overwhelming majority of microscopic configurations.[9] And from this definition follows that of the irreversible evolution toward equilibrium. If we assume the existence of a state of low probability brought about by a small number of microscopic configurations, the most likely path should lead to a more probable state, to which corresponds a greater number of elementary configurations. In other words, the growth of entropy does not tell us anything about a physical change over time—this falls within the jurisdiction of dynamics—it only tells us about the change over time of the system as we are able to observe it, the macroscopic system each of whose states can correspond indifferently to a multitude of distinct dynamic states.

To claim that irreversible thermodynamic change corresponds to the most likely change is convenient and reassuring. But only for someone who wants to settle, once and for all, a question of rights, of the distribution of responsibilities. If it's a question of productivity, Boltzmann's kinetic equation remains incomparable: it and similar equations gave meaning to the experimental values that characterized nonequilibrium behavior.[10] Probabilities couldn't be used to discuss such behavior any more than the older thermodynamic potential. Moreover, and the fact that this remarkable characteristic is rarely emphasized reflects the process of percolation—history flows through some problems and not through others—we cannot say that probabilities explain the difference between the changes predicted by the second law and those it prohibits. On the contrary, the reversal of velocities allows us to systematically correlate to each particular dynamic state involving a change that corresponds to an approach to macroscopic equilibrium another state that will bring about a change corresponding to a spontaneous movement away from that equilibrium. We must conclude, therefore, that the a priori probability of the two kinds of change *is the same.* "Irreversible" thermodynamic change is not favored by the probability argument over the change prohibited by the second law. A similar difficulty is found in the definition of probabilities themselves. Given a macroscopic nonequilibrium—that is, improbable—state, probability allows us to predict that there will be a change to a more probable state, one closer to equilibrium. But the same argument would also lead us to conclude that this "improbable" state is, according to probability, the result of a "more probable" state situated in the past. The argument based on probability is, therefore, implicitly limited to the calculation of future probabilities. It doesn't explain irreversibility; it simply makes it compatible with dynamic reversibility. It can only convince those who want to be convinced.[11]

This was the objection made in 1896 by Zermelo, who revealed that irreversible change was not favored. Zermelo used a dynamic theorem by Poincaré, according to which any dynamic system, and therefore (and why not?) the universe, always returns to some point arbitrarily close to its initial state. In this case, probabilities are powerless, because it is not a question of preparing an "improbable" state but of a spontaneous dynamic change. And Boltzmann, once again, backed off. He recognized that, at the scale of the universe, which we can conceive of as being overall at equilibrium, there exist as many regions where it is moving away from equilibrium as there are regions, like our own, where equilibrium belongs to the future. The second law of thermodynamics is entirely relative to a *cosmological contingency,* to the fact that our region of the universe constitutes an improbable local fluctuation, a deviation from equilibrium that is in the process of being neutralized.

20

In Place of an Epilogue

Zermelo's proof was directed at a dynamic system. Boltzmann recognized that the universe itself was incapable of serving as a witness to the difference between past and future. So he gave up trying to promote anything that might separate the two, as if it were self-evident that the universe satisfied the requirements of dynamics and fell within the jurisdiction of the Queen of Heaven. Even today, those who accept the probabilistic interpretation of entropy—the great majority of physicists—acknowledge that, in one way or another, the Sun doesn't really burn, and that there must exist a point of view, the only true point of view, that would allow us to relate the solar furnace to some "purely macroscopic" effect. On Earth, as in heaven, "nothing happens," nothing other than the tranquil, unchanging repetition of a reality wholly subject to the power of the = sign.

What has happened? Why was the probabilistic interpretation accepted and why is it still accepted today? How did we silently move from the problem posed by the degradation of energy, the spontaneous smoothing of differences during the passage toward thermodynamic equilibrium, to the most audacious negation ever invented, the negation of the "arrow of time"? And why, we must again ask, weren't the variety of kinetic events, which, with the exception of elastic collision, all seem to point to a difference between past and future, used to help resist the power of dynamics?

With respect to Boltzmann, we can invoke the continuation of the operation of capture, when the qualitative multiplicity of energy transformations was unified by a single question, as defined in the arena constituted by the Carnot cycle. The transformations between heat and work are among those where the only type of event required by kinetic models is the "collision," and it was with respect to such transformations that the law of increasing entropy was first discussed. We can then say that, for Boltzmann, it was here that the relationship between dynamics and thermodynamics was fully articulated. Just as the various forms of energy had been captured by the contrast between conservation in the mechanical sense, on the one hand, and conservation *and* degradation in the thermodynamic sense, on the other, the variety of kinetic events could be said to have been captured by an issue that had caused a crisis in physics: was the physicist obligated to rely solely on observable phenomena as characterized by thermodynamics or was she free to construct models involving unobservable actors, "beyond phenomena"? To endow those actors with "dissipative" properties that make them, individually, witnesses of the arrow of time would have meant, in this context, relying on some "deus ex machina." It would have meant assigning the responsibility for what the model was specifically supposed to bring about—thermodynamic reversibility—to unobservable actors. Such a capture effect is fairly commonplace. Rather than changing the problem, especially if there is considerable prestige associated with it, a researcher will often present the impasse she has arrived at as an unavoidable conclusion, a dizzying prospect for us all.

However, although this capture effect may hold true for Boltzmann, it doesn't help us understand the success of the probabilistic interpretation. Even more so given that at the time of Boltzmann's death, "percolation" had not yet taken place. Boltzmann himself had been on the defensive until the end, and, in Vienna, his adversary Mach, the anti-atomist, continued to hold the upper hand. Another "history" still seemed possible, one in which "atoms" justified a kinetic approach but escaped the jurisdiction of the Queen of Heaven. The question gains in relevance given that those who studied these unobservable actors at the turn of the century were trying out new experimental techniques involving increasingly diverse "events"—radioactive decay, the emission and absorption of light, and so on—and were far from being worshippers of the Queen of Heaven. Each of them was, in fact, more or less convinced that going beyond phenomena meant going beyond dynamics as well as thermodynamics. Take Jean Perrin's great work, *Atoms,* which, in 1912, celebrated the event that gave atoms a preeminent position in physics: we can count them, we can assign a value to Avogadro's number. For Perrin, the world of atoms signifies the destitution of regular laws. For it is the mad, ever-interrupted, and ungoverned race of the Brownian particle, no longer that of the planets in the heavens, that now provides its pertinent figure to motion. Consider the first "quantum" model of the atom, that of Niels Bohr, which contradicted, explicitly and deliberately, the laws of dynamics and electrodynamics. There are other examples as well, such as Paul Langevin, who, in 1904, attempted to reduce the physics of mass and inertia to a particular case in the more general framework of the nonmechanical representation of matter.[1] What meaning did Einstein give to the theory of special relativity in 1905?[2] It seemed obvious to many that the limits of dynamic intelligibility had been reached: to go beyond phenomena also meant going beyond the old laws of motion. Determinism itself was violently attacked in the name of the exponential character of the spontaneous decay of unstable atomic nuclei, and Oswald Spengler, in *The Decline of the West,* praised the return of "freedom" to the heart of physics.

How can we tell a story without an epilogue? The epilogue is usually the point in the narrative when all the ingredients have played their part and can, one last time, be brought together and viewed in terms of their final role in the story. It is the moment when the events that have marked the narrative can be seen by the various protagonists in terms of the contrast between the multiple meanings they may have been assigned and the meaning they finally embody. But the singularity of our story, its irony, is that it ends in the hands of those who consider it already over, because it no longer holds any interest for them. The triumph of the Queen of Heaven, before whose power Boltzmann bowed down, never really took place. The arena in which the "anti-atomists" of the nineteenth century and the partisans of the kinetic hypothesis did battle, the question of the difference between mechanical and thermodynamic equilibria, has simply been abandoned. The question of the kinetic event itself has found no representative because, wrapped in the question of the status of macroscopic dissipation, it has shared its fate, it has become part of a story that is no longer relevant.

The Queen of Heaven did not triumph because, for the new generation of physicists who looked beyond phenomena toward atoms, nuclei, electrons, and ions, it was nothing more than a ragged shadow. And yet, the probabilistic interpretation triumphed even though its original function had been to subject dissipative macroscopic phenomena to the jurisdiction of the equations of dynamics. Maxwell's demon resisted the fall of the dominion of dynamics. Why?

At this point we may have to take into account a last, unforeseeable, ironic connection that finally triggered the crossing of the threshold of percolation. Probabilities may have triumphed because of their association, from the very first years of the twentieth century, with a turning point in physics, with an episode that cast the old quarrels between dynamics and thermodynamics back into prehistory: the creation of the first nonmechanistic *theory* referring to a reality "beyond phenomena." I'm referring to Max Planck's quantum theory of light absorption and emission. It is this theory that marked the collapse of Hamiltonian dynamics as a fundamental mode of intelligibility when it described "black-body radiation" in terms of the discrete distribution of energy. But it is the term "distribution" that should draw our attention. The discrete character of "light quanta" is based on an argument that makes use of statistical distribution and, therefore, probabilistic reasoning. Consequently, it is not impossible that the probabilistic argument was accepted not as a result of the triumph of the Queen of Heaven, but because probabilities had become an instrument of cutting-edge physics. Thomas Kuhn makes an important point in this regard. Whereas in 1900 the number of physicists using probability could be counted on both hands, the discovery of quantum discontinuity would make it a required instrument, one that was crucial for any physicist planning to tackle the new reality.[3] From that point on, the ability of probability to interpret a former reality no longer of interest to anyone, the irreversible dissipation of energy, would have seemed self-evident.

If this is the case, the conclusion of the history whose anamnesis I have tried to develop is marked by a profound irony. Planck's work on black-body radiation, in 1894, began with a return to Boltzmann's attempt to provide a physical interpretation for irreversibility. Planck was a thermodynamicist. He was wholly engaged by the distinction between mechanical and thermodynamic potentials, and fascinated by the possibility of providing an objective description for irreversible processes. Faced with Boltzmann's failure, he concluded that elastic collisions among particles might be the wrong place to look and that another, more promising, case would allow for a rigorous explanation, without approximations this time, of the approach to equilibrium, together with a physical interpretation of the process. This case was none other than the equilibrium of the radiation in a black body. The light absorbed and reemitted from a cavity with opaque walls underwent an "irreversible" change during which it "forgot" its initial energy distribution and became "black-body radiation," whose energy distribution was a function of the temperature of the cavity alone. Black-body radiation was a disappointment to Planck. The laws of electrodynamics drove him, as the laws of dynamics had driven Boltzmann, to an interpretation of the approach to equilibrium that was "merely probabilistic." But it was when he had to connect those unfortunate probabilities to experimental results (which concerned the energy distribution at equilibrium of the radiation of a black body as a function of temperature) that something unexpected happened: only the hypothesis of a discontinuous, or "quantum," distribution of radiant energy could account for the facts.

There is no epilogue to this story, because its subject has changed its identity.[4] As the protagonists of nineteenth-century physics were trying, each in their own way, to understand what it was that experimental relations required and the consequent obligations for the physical–mathematical edifice, which would claim to articulate them coherently, a new protagonist now appeared on the scene, *theoretical physics.* This physics was to herald, as Einstein had, the *enigma* of the intelligibility of the world. In other words, it honored "enigmatic factishes," but—contrary to Duhem—claimed that their enigma can be brought to light and that our ability to do so is the only true enigma.

Theoretical physics is, by definition, revolutionary because it is explicitly associated with the thesis of revolutionary transformation, as illustrated by the theory of light quanta and relativity. Which is a way of saying that it also introduced new values and obligations. It no longer celebrates its roots in experimentation and its corresponding obligations. This is not to say that it had contempt for "facts," quite the contrary, but that, compared to them, it claimed a new freedom that allowed it to designate what was beyond those facts as the only worthwhile challenge. Its first obligation was not so much to get beyond phenomena—that was the authorizing precedent, the solved enigma, the acquired knowledge whose heritage it celebrated—as to turn that beyond into a revolutionary resource whose value would be measured in terms of the denials of phenomenological "evidence" it will allow to be made. Enigmatically but incontestably, the intelligibility of the world must make itself known, and the sign that it has done so is the wound it inflicts in our evidence.

To these new values of physics correspond new requirements addressed to the world and to humankind. The new identity of physics passionately requires that the world justify, and that humans accept, physicists' right to freely negotiate how and to what extent obligations bind them to the phenomena beyond which their vocation demands they venture. The way in which probabilistic interpretation resolves the question of dissipative change, and with it that of the diversified body of physical–chemical phenomena, has, from the point of view of these revolutionary values, become exemplary and deeply satisfying: it reflects the physicist's freedom not to take "observable phenomena" into account but to *discount* them.

If I were a historian of science, many things would remain to be said, primarily the differentiation of time that characterizes the conclusion of this episode of nineteenth-century physics. "French" science, for example, resisted the new revolutionary values longer than others and was thus judged to be "lagging behind." But anamnesis can come to a close, for what follows lives in the experience of all of us. Each of us has learned to associate the triumph of physics with the shocking discovery of the illusory nature of what we thought we knew. Does Schrödinger's cat die before the observer opens the box? Anyone who accepts that physics is likely to ask this kind of question (which I will discuss in "Quantum Mechanics: The End of the Dream" in *Cosmopolitics II*) is prepared to accept anything from this revolutionary science. And anamnesis will have done its job if it has succeeded in destabilizing what characterizes the history of twentieth-century physics, the way in which it is presented to us, as well as the way in which it is present to itself: the enthusiastic disqualification of "common knowledge" in the name of the "revolutions" physics imposes on us all and the pathos associated with the great theme of the "vocation of the physicist," who aspires to a unified conception of the world beyond heterogeneous empirical phenomena. The hierarchy that corresponds to this vocation within physics itself is not the outcome of some gradual clarification whose logic we could understand, but translates a brutal fact: "phenomena" can be subordinated because those who are interested in them are themselves subordinate, left behind by revolutionary physics.

NOTES

PREFACE

1. [The present translation is based on the updated two-volume French edition of *Cosmopolitics* published in 2003. The contents of that two-volume edition were compiled from an earlier edition of *Cosmopolitics* published in 1997 in seven volumes. *Cosmopolitics I* includes volumes I, II, and III of the original edition; *Cosmopolitics II* includes volumes IV, V, VI, and VII of the original edition. Volume references throughout this text follow this numbering scheme. For example, Book IV refers to volume 4 of the original edition, which is included in volume 2 of the present edition. Please note that the current English-language edition of *Cosmopolitics* has been revised by the author and varies slightly from the 2003 French edition. All translations of quoted French works are my own, except where otherwise noted. —*Trans.*]

1. SCIENTIFIC PASSIONS

1. See *D'une science à l'autre: Les concepts nomades,* ed. Isabelle Stengers (Paris: Le Seuil, 1987).

2. Ilya Prigogine and Isabelle Stengers, *Order out of Chaos: Man's New Dialogue with Nature* (New York: Random House, 1984). However, the quotation, which is found in the conclusion of the French edition of *Order out of Chaos* (*La Nouvelle Alliance*), does not appear in the English translation. See Isabelle Stengers, *Power and Invention* (Minneapolis: University of Minnesota Press, 1997), 42.

3. Ilya Prigogine and Isabelle Stengers, *Entre le temps et l'éternité* (Paris: Flammarion, 1992), 193–94.

4. Max Planck, "Die Einheit des Physikalischen Weltbildes," *Physikalische Zeitschrift* 10, 62–75, reproduced in *Physical Reality,* ed. S. Toulmin (New York: Harper Torchbooks, 1970).

5. In Gilles Deleuze and Félix Guattari, *What Is Philosophy?,* trans. Hugh Tomlinson and Graham Burchell (New York: Columbia University Press, 1994), 66–68, the distinction between "psychosocial type" and "conceptual personae" ("psychosocial types . . . become susceptible to a determination purely of thinking and of thought that wrests from them both the historical state of affairs of a society and the lived experience of individuals," 70) refers to the distinction between history and event. Every psychosocial

type—Greek, capitalist, proletarian, and so on—makes perceptible the corresponding territory it establishes, its vectors of deterritorialization, its processes of reterritorialization. The compound adjective—"psycho" and "social"—indicates that the type is relative to a given society, a given historical moment: it is only under these conditions that such adjectives have meaning.

6. Mach is only an example. However, some may recall Bergson's criticism of Einstein's theory of relativity, as well as, in the field of quantum mechanics, the way in which the proponents of the Copenhagen interpretation fought against the "positivist curse," that is, against the presentation of their hypotheses as an illustration of empirico–critical doctrines (see Book IV, "Quantum Mechanics: The End of the Dream").

7. Isabelle Stengers, *The Invention of Modern Science,* trans. Daniel W. Smith (Minneapolis: University of Minnesota Press, 2000), 145. My approach to capitalism is not unrelated to that found in *Anti-Oedipus: Capitalism and Schizophrenia* by Gilles Deleuze and Félix Guattari, trans. Robert Hurley, Mark Seem, and Helen R. Lane (Minneapolis: University of Minnesota Press, 1983), which illustrates capitalism's radical indifference to values such as those associated with "modernity." The movement of capitalist deterritorialization has as its correlate continued operations of reterritorialization: the resurrection of "new archaisms" or the maintenance of old territorialities redefined by new forms of coordination-conjunction (for example, family territory is redefined by its conjunction with the household appliances industries, just as the "protection of nature" can be redefined by its conjunction with the "green" label).

8. Deleuze and Guattari, *What Is Philosophy?*, 113.

9. For a remarkable presentation of this obviously crucial distinction, see François Zourabichvili, *Deleuze: Une philosophie de l'événement* (Paris: Presses Universitaires de France, 1994).

10. I am giving to the word *possible* the meaning frequently given by Deleuze to *virtual*. To create a more explicit connection with scientific practices, I have decided to use the term "probable" for the Deleuzian "possible," which lacks only reality. The calculation of probabilities assumes, like all calculations, the conservation of whatever the calculation has been constructed from. That is why probability commits whoever makes reference to it to maintaining that conservation. Naturally, this commitment assumes very different meanings depending on whether it reflects the creation of a measurement (the rate of disintegration of radioactive nuclei, for example), the risk of decision making, or the claim to a "realist" vision of things.

2. THE NEUTRINO'S PARADOXICAL MODE OF EXISTENCE

1. Émile Meyerson, *Identity and Reality,* trans. Kate Loewenberg (London: Gordon and Breach, 1989), 230–31. Here, Meyerson appears to be

anticipating Einstein's general theory of relativity. Moreover, he has recognized the "acosmic" vertigo brought on by the four-dimensional cosmos of general relativity. *Relativistic Deduction* (1925, 1989) is the first exposé of Einstein's work that does not focus on special relativity and the operationalist arguments supporting it, and directly addresses Einstein the "metaphysician," who assumed such importance to his surprised colleagues and embarrassed positivist philosophers. Similarly, in *Réel et déterminisme dans la physique quantique* (Paris: Hermann, 1933), Meyerson predicted that the painful denial quantum physics has forced physicists to accept may very well be unstable: "There can be little doubt that if there were the slightest possibility, researchers would quickly return to a somewhat concrete image, realizable in thought, of the universe, a *Weltbild* to use Planck's expression" (49). I emphasize this point because the Bachelardian philosophical tradition in France has followed Bachelard in a systematic counterreading of Meyerson, who, according to Dominque Lecourt, for example, embodies "that pretentious philosophy that glorifies the longevity of its questions and wants to subject scientific knowledge to its decrees" (*L'Épistémologie historique de Gaston Bachelard* [Paris: Vrin, 1974], 34). This should be compared with Einstein's correspondence with Meyerson, published in Albert Einstein, *Œuvres choisies*, vol. 4, *Correspondances françaises* (Paris: Seuil-CNRS, 1989). Even Lenin, in *Materialism and Empirio-Criticism*, treated his adversaries better than Bachelard and his emulators have treated Meyerson.

2. Meyerson, *Identity and Reality*, 377.

3. Bruno Latour, *Petite Réflexion sur le culte moderne des dieux faitiches* (Le Plessis-Robinson: Synthélabo, "Les Empêcheurs de penser en rond," 1996).

4. This construction is ongoing. The question of knowing whether neutrinos have mass and therefore of knowing "what" we are detecting is still open, or was until 2002 when the Nobel Prize for physics was awarded to three physicists for determining that neutrinos are not massless.

5. For a different approach to the same distinction, see Rom Harre, *The Principle of Scientific Thinking* (London: Macmillan, 1970), and Roy Bhaskar, *A Realist Theory of Science* (Leeds: Leeds Books, 1975).

6. Latour, *Petite Réflexion sur le culte moderne des dieux faitiches*, 99.

7. For the theme of the event considered in terms of scientific coordinates, see Isabelle Stengers, *The Invention of Modern Science*, trans. Daniel W. Smith (Minneapolis: University of Minnesota Press, 2000).

3. CULTURING THE *PHARMAKON*?

1. See the magisterial text by Jacques Derrida, "Plato's Pharmacy," in *Dissemination*, trans. Barbara Johnson (Chicago: University of Chicago Press, 1983). However, I do not necessarily accept the point of view proposed,

where the multiplicity of *pharmaka* is subtly channeled toward the overarching question of writing.

2. Tobie Nathan, *L'influence qui guérit* (Paris: Éditions Odile Jacob, 1994), 29.

3. A disparate series, as its unity is found only in the fact that its members function as a kind of foil. Its number includes David Hume, whose critique of the concept of causality awakened Kant from his "dogmatic sleep," and prompted him to search for a foundation that would allow us to forget Humeian "habits," Henri Bergson, who was accused of reducing human freedom to that of an apple, the American pragmatists at the beginning of the twentieth century, accused of seeing no other justification for values than as a kind of utilitarian calculus, and Michel Foucault, denounced by the real "philosophers of communication" for not claiming to establish the universal validity of his political commitment. That's just a sample. The weakness of this series is the monotony of the accusation and the perspective it leads to: each author is asked to justify the very thing he had taken the risk to create. On the other hand, this series is a good illustration of the difference pointed out by Deleuze and Guattari between "majority" and "minority." Here, the majority can always appeal to the good sense of everyone else and function as a "restoring force" in the history of philosophy: a restoring force in the mechanical sense, where it pulls the stretched spring back to its initial position of equilibrium, and in the memorial sense, where it reenacts the primitive scene of the exclusion of the sophists.

4. See Léon Chertok, *L'Énigme de la relation au cœur de la médecine*, ed. and introd. Isabelle Stengers (Le Plessis-Robinson: Delagrange/Synthélabo, 1992).

5. See Tobie Nathan and Isabelle Stengers, *Médecins et sorciers* (Le Plessis-Robinson: Synthélabo, "Les Empêcheurs de penser en rond," 1995).

6. Isabelle Stengers, *The Invention of Modern Science*, trans. Daniel W. Smith (Minneapolis: University of Minnesota Press, 2000), 89.

7. In *Demystifying Mentalities* (Cambridge: Cambridge University Press, 1990), Geoffrey Lloyd illustrates the practical irreversibility created by the enunciation of explicit categories, such as those that refer to magic, or metaphor, or the *pharmakon:* "The questions, then, of how the actors themselves perceive their own activity, or the conventions within which it fits or from which it deviates, the traditions that do or do not sanction it, are prior to and independent of the question of the existence of some such category as magic itself. But once that category exists, it can hardly fail to change the perception. . . . for the category enabled the challenge to justify the activity to be pressed. . . . and the activity could no longer remain, or could not do so easily, an unquestioned item invisible—or indistinctive—against the background of the traditions to which it belonged" (69).

8. No doubt, it would be appropriate to speak of "eco-ethology" to distinguish the ecology I'm speaking of from "systemic" or "economic" ecology, where "large-scale" functional relationships and assessments are the norm. But this would mean accepting as well-founded a distinction that I see no reason to maintain: the ecology of systems and assessments do not deserve the honor.

9. "Bats Sow Seed of Rainforest Recovery," *New Scientist,* no. 1930 (June 18, 1994): 10.

10. See Bruno Latour, "Moderniser ou écologiser? À la recherche de la 'septième cité,'" *Écologie politique,* no. 13 (1995): 5–27.

11. Deleuze and Guattari have introduced the concept of "double capture," the typical illustration of which, the wasp and the orchid, also refers to the field of evolution. In his *Dialogues* with Claire Parnet (New York: Columbia University Press, 1977), Deleuze gives this concept, which views becoming as a kind of event, its fullest scope: a-parallel evolution, unnatural marriages binding two kingdoms, or, according to the ethologist Rémy Chauvin whom he quotes, "two beings that have nothing whatsoever to do with one another" (2–3). The concept of "double capture" has a much broader scope than the "reciprocal capture" I am using here. It views all relationships as event and is, therefore, relevant for describing what occurs in the "interview" between Deleuze and Parnet, or in any situation where one might be tempted to speak of an "exchange." "Reciprocal capture" itself refers to a double capture that creates a relationship endowed with a certain stability. It is relevant whenever the "marriage" produces, as is the case for the wasp and the orchid, but not necessarily for an interview, identifiable heirs among whom we are tempted to distribute—or that will be distributed among themselves—the attributes that explain the relationship and justify its stability.

12. On the other hand, the caterpillar specialist who learns to recognize them in spite of their "disguise" incorporates the production of this elaborate perceptive capacity in the construction of her professional identity. As for the caterpillars, they suffer the consequences of this ability to identify them: the specialist exists "for" the caterpillars, affirms their existence through her own, but the opposite is not true.

13. Félix Guattari, *Chaosmosis: An Ethico-Aesthetic Paradigm* (Bloomington: Indiana University Press, 1995), 47 and 33.

14. To avoid introducing traditional philosophical references (Nietzsche, Whitehead), it would be useful to recall Popper's distinction between the first world, in which we have not until now needed to recognize the connection between risk and existence, and the second world, where that distinction is essential.

15. See Gilles Deleuze and Félix Guattari, *What Is Philosophy?,* trans. Hugh Tomlinson and Graham Burchell (New York: Columbia University Press, 1994), as well as *L'Effet Whitehead,* ed. Isabelle Stengers (Paris: Vrin, 1994).

16. See Stengers, *The Invention of Modern Science*, 48–53.

17. See Pierre Lévy, *Les Technologies de l'intelligence* (Paris: La Découverte, 1990).

4. CONSTRAINTS

1. See *La Science et ses réseaux: Genèse et circulation des faits scientifiques*, ed. Michel Callon (Paris: La Découverte, 1989), and, of course, Bruno Latour, *Science in Action: How to Follow Scientists and Engineers through Society* (Cambridge, Mass.: Harvard University Press, 1988), and *Aramis, or the Love of Technology*, trans. Catherine Porter (Cambridge, Mass.: Harvard University Press, 1996).

2. This is Gérard Fourez's conclusion in *Alphabétisation scientifique et technique: Essai sur les finalités de l'enseignement des sciences* (Brussels: De Boeck University, 1994). But the discussion that leads to this conclusion introduces some very interesting problems. For example, one wonders whether it isn't within the framework of "technological" education that the abstract "results" of the sciences (the "laws" of physics or chemistry detached from their histories) assume meaning. Within this framework they appear as "constraints," an obstacle to be overcome or part of a solution. They (the sciences) assume a meaning and importance that are effectively detachable from their scientific mode of production because they are redefined by other attachments.

3. Here, it is useful to recall that it was obviously not biologists who invented the question of our "ancestors," and that the idea that our "true" ancestors were hominids does not at all correspond to the way the problem of our ancestors has been raised by different peoples on Earth. How can we avoid presenting this new "genealogy" as being "truer" than theirs for all the humans it claims to affect? How can it be "present" without imposing a dissociation between what is nothing but a mythical story, culturally respectable but not much more, and objective scientific knowledge? It is because we don't know the answer to this that the ecology of practices today is speculative. In fact, the creationist controversy in the United States might lead some to criticize the danger of such speculation. Aren't I providing arguments for their cause and for others that are certainly going to arise? We should be wary of the apparent limpidity of the American creationist "case," of the ease with which "good" and "bad" are identified. If American creationists are disturbing, it is primarily because they position themselves as representatives of an authority that they oppose to that of the sciences and whose effects we have every reason to fear. It is the great strength of modernity to generate caricatural, monstrous, heinous oppositions, against which no hesitation is possible. Moreover, ever since the Supreme Court's decision that creationism was a religious idea and not a scientific statement, conflicts have continued to erupt locally, and they are not without interest. The

creationists now want to force teachers to *discuss* the theory of evolution, to point out its possible weaknesses, to present alternatives without disqualifying them a priori. Naturally, we cannot ignore the activities of parents' committees and other pressure groups. Nonetheless, it remains that this retreat and the turbulence and difficulties it has caused, are symptomatic. It is as if collectives were needed, capable of providing organized resistance, tenacious and fanatic, to certain types of scientific knowledge, so that the transmission of that knowledge in schools might acknowledge its risky, selective, interesting mode of existence—the very thing that demonstrates its scientific nature.

4. See François Gros, *Les Secrets du gène* (Paris: Éditions Odile Jacob, 1986).

5. Bruno Latour, *We Have Never Been Modern,* trans. Catherine Porter (Cambridge, Mass.: Harvard University Press, 1993),115.

6. Luc Boltanski and Laurent Thévenot, *On Justification: Economies of Worth,* trans. Catherine Porter (Princeton, N.J.: Princeton University Press, 2006).

7. Isabelle Stengers, *The Invention of Modern Science,* trans. Daniel W. Smith (Minneapolis: University of Minnesota Press, 2000), 89.

8. I am obligated to Bernadette Bensaude-Vincent for the comment and, for this and many other reasons, owe her a debt of gratitude.

9. Is mathematics a modern practice? It's an apparently absurd question if we recall the role that mathematics plays in modern knowledge as well as its obvious collusion with power. Yet, the concepts of requirement and obligation raise several questions. First, and in the context of the presentation I am proposing here, mathematics is obviously not the heir of the Platonic gesture of excluding the sophists. Rather, it is the example of mathematics that Plato requires to bring about a truth that is not dependent on opinion. Also, the collusion between mathematics and power is not a capitalist invention: geometry, astronomy, and arithmetic have always been an integral part of the material and spiritual management of empire. Finally, the polemical relationship that exists among the modern sciences does not exist between mathematics and those same modern sciences. If we had to situate it in the hierarchy, the only suitable place would be the summit, but this position obscures its singular mobility: the mathematical future can, for better and for worse, affect any science, without, however, creating a dependent relationship, although, wherever it is *transported,* wherever *relationships* arise, mathematical reason orders, commands, and declaims its law (see Michel Serres, *Origins of Geometry* [Manchester: Clinamen Press, 2002]).

Here I want to limit myself to pointing out that the concepts of obligation and requirement can help us grasp the singularity of mathematics. No doubt, it is the practice in which those concepts are most obviously prevalent. Mathematical beings exist only to the extent that they satisfy a requirement that their definition makes explicit. This requirement must resist all challenges, remain unchanged under all circumstances. Correlatively, its definition obligates the mathematician, constrains her to the most perilous forms of invention, forces her to confront what others would consider unthinkable. It is quite possible that the obligation to introduce irrational numbers did not inspire the disciples of Pythagoras with panic as we are told, but the myth is appropriate to mathematics. And it may be the fact that, in mathematics, requirement and obligation are literally and inseparably constitutive of what exists and of what causes to exist, that it is a requirement of every definition that it be followed to its most scandalous consequences, which confers singularity on this practice when compared to those I call modern. These have to invent, in fields that are always already inhabited by others, the means for creating a difference between a reliable statement and simple opinion, that is, the means to "exclude the sophists," whereas the mathematician brings into existence conceptual spaces that no one can inhabit without accepting their constraints.

5. INTRODUCTIONS

1. Gilles Deleuze, *Spinoza: Practical Philosophy* (San Francisco: City Lights, 2001), 27.

2. Ibid., 23.

3. Gilles Deleuze and Félix Guattari, *What Is Philosophy?*, trans. Hugh Tomlinson and Graham Burchell (New York: Columbia University Press, 1994), 34.

4. Ibid.

5. Ibid., 159–61.

6. The reference is to Charles Ferdinand Nothomb (July 1996). For one approach to such uncontrollable slippage, see Isabelle Stengers and Olivier Ralet, *Drogues: Le défi hollandais* (Le Plessis-Robinson: Éditions Delagrange/Synthélabo, "Les Empêcheurs de penser en rond," 1992).

7. In *What Is Philosophy?* we find the description-creation of a philosophical experience that actively excludes any possibility of "generalist" knowledge, erasing the differences between philosophy, art, and science. This example is sufficient to show that the experience of "here" is not at all synonymous with a ban on speaking about what is happening "elsewhere." Rather, it's a question of speaking of here, in this case, of creating *concepts* that bring into existence the question of art and science, not of ruling on the respective domains of philosophy, art, and science.

8. It may also have situated Prigogine, but it's not up to me to report it.

9. For example, the first appendix to the French edition of *Order out of Chaos* assigns a key role to the strictly philosophical argument that a difference in kind exists between knowledge whose precision is positively finite and knowledge whose precision is unlimited, tending toward infin-

ity. This argument, which was used to claim that every physical description had to be strong in terms of approximation, provided dynamic chaos theory with the power to challenge the legitimacy of classical laws. It became secondary, or propaedeutic, in Ilya Prigogine's *The End of Certainty* (New York: Free Press, 1997), and was even criticized as inadequate (223), as something ancillary—*and therefore weak*—in a line of argument that should be purely physical. In the interim, dynamic chaos theory had given the green light to another mode of argument.

10. On the subject of chemistry and its contemporary singularity—a science that no longer dominates anywhere in spite of the fact that it is "everywhere"—see Bernadette Bensaude-Vincent and Isabelle Stengers, *The History of Chemistry*, trans. Deborah van Dam (Cambridge, Mass.: Harvard University Press, 1996), 97.

11. I don't know to what extent the "regimes of justification" of Boltanski and Thévenot can renew the practices of the social sciences, but reading *On Justification: Economies of Worth* can at least, through the sophistication of interregime negotiation described there, contribute to the creation of a new kind of manual for living, where we would introduce ourselves to others in a way that is not exotic or ironic—for we are indeed attached to those regimes of justification—but humorous. The singularity of "worth" that we cultivate implies and seeks the possibility of other cultures.

12. The analogy is based on the fact that the chemical identity of oxygen gas is assumed not to undergo transformation by this capture. Obviously, the same is not true for the multiple chemical transformations in which oxygen molecules are captured following the "making available" that invented them as resources for the living.

13. See Léon Chertok and Isabelle Stengers, *Le Cœur et la Raison* (Paris: Payot, 1989), and Isabelle Stengers, *La Volonté de faire science: À propos de la psychanalyse* (Le Plessis-Robinson: Delagrange/Synthélabo, "Les Empêcheurs de penser en rond," 1992).

6. THE QUESTION OF UNKNOWNS

1. Gilles Deleuze, *Difference and Repetition*, trans. Paul Patton (New York: Columbia University Press, 1995), xxi.

2. See Gilles Deleuze and Félix Guattari, *What Is Philosophy?*, trans. Hugh Tomlinson and Graham Burchell (New York: Columbia University Press, 1994), 2: "But the answer not only had to take note of the question, it had to determine its moment, its occasion and circumstances, its landscapes and personae, its conditions and unknowns."

3. I am thinking especially of mathematical economics (focused on the notion of utility). The field embodies the worst kind of duplicitous language. "Obligations" are entirely determined by the ability of language to construct and by the theorems this language is likely to generate—some-

thing economists are, whenever the circumstances allow, even more willing to recognize because it allows them to reject "external" criticisms. Their theory demands nothing of the world, it is self-referential. But under other circumstances the majority of them maintain a prudent silence whenever these models are used to justify certain economic policies and thereby become vectors of literally obscene judgments.

4. See Gilles Deleuze and Félix Guattari, *Anti-Oedipus: Capitalism and Schizophrenia,* trans. Robert Hurley, Mark Seem, and Helen R. Lane (Minneapolis: University of Minnesota Press, 1983). To claim that capitalism is not a practice also means that it is radically distinct from the practices of those who operate within the coordinates it defines and redefines from era to era. Correlatively, Marx's denunciation of the fetishization of goods retains its relevance, providing we add that it is not a general critique of fetishes. Those clever souls who claim that "the" commodity doesn't exist because every commodity presents quite specific practical problems, might just as well claim that "the" capitalism doesn't exist either because they have never run into it. Obviously, we never encounter "the" commodity or "the" capitalism as long as we fail to adopt, like Deleuze and Guattari, the "indispensable" position of the "incompetent," someone who dares assert that all this history, bursting with competent strategies and practices, "is profoundly schizoid" (238).

5. Sigmund Freud, "Analysis, Terminable and Interminable," in *The Standard Edition of the Complete Psychological Works of Sigmund Freud,* trans. James Strachey, in collaboration with Anna Freud, assisted by Alix Strachey and Alan Tyson (London: Hogarth Press and the Institute of Psycho-Analysis, 1966), 23:209–53.

6. Obviously, Freud was referring only to psychotherapeutic techniques, but it is these techniques that, in fact, reflect all the dimensions of the question of healing. So-called modern medicine is all the more powerful in that it can shift the problem and, for example, overcome the bacteria that proliferate in a body rather than "heal" that body. See Isabelle Stengers, "Le médecin et le charlatan," in Tobie Nathan and Isabelle Stengers, *Médecins et sorciers* (Le Plessis-Robinson: Synthélabo, "Les Empêcheurs de penser en rond," 1995).

7. It is significant that Freud writes about the three impossible professions at the moment he recognizes the limitations of therapeutic technique offered by psychoanalysis but refuses to disavow the requirements and obligations that allow it to exist as a science. If psychoanalysis had satisfied the hopes of its creator, it might have avoided disqualifying other therapeutic practices the way physics might have avoided proffering visions of the world in whose name all other forms of knowledge are disqualified. It is in terms of the practical failure of the requirements that psychoanalysis makes of the human psyche, that is, in terms of the disappointment of the practi-

tioners of psychoanalysis, that its history should be considered. See Isabelle Stengers, "Les déceptions du pouvoir," in *Hypnose, influence, transe,* ed. Daniel Bougnoux (Paris: Éditions Delagrange, "Les Empêcheurs de penser en rond," 1991), 215–31, and *La Volonté de faire science: À propos de la psychanalyse* (Le Plessis-Robinson: Delagrange/Synthélabo, "Les Empêcheurs de penser en rond," 1992).

8. Freud, for example, essentially believed that unconscious conflicts preexisted, in one way or another, the therapeutic technique that revealed their existence.

9. Alexander Pope (*Essays on Criticism,* 1711): "Fools rush in where angels fear to tread." The fragments written by Gregory Bateson were published after his death by his daughter Mary Catherine Bateson as *Angels Fear: Towards an Epistemology of the Sacred* (New York: Bantam Books, 1988).

10. Bateson and Bateson, *Angels Fear*, 64.

11. Bruno Latour, *We Have Never Been Modern,* trans. Catherine Porter (Cambridge, Mass.: Harvard University Press, 1993), 33. Naturally, the premodern Christian god does not have, at least not in my use of the quote, any a priori privilege compared to the gods, ancestors, and spirits who haunt the "ancient *phusis*" and the "old anthropological collective."

12. Bruno Latour, *Irréduction* (published with *Les Microbes: Guerre et paix*) (Paris: Anne-Marie Métaillié, 1984), 232.

13. See *Perpetual Peace* (1795) and *Anthropology from a Pragmatic Point of View* (1798). For Kant, peace, in the strictly cosmopolitical sense, derives from culture rather than the conscious intent of individuals—morality. This distinction serves my purpose, although I give it a constructivist translation: I am interested in practices and their corresponding psychosocial types, rather than individuals.

14. Bruno Latour, *The Pasteurization of France,* trans. Alan Sheridan and John Law (Cambridge, Mass.: Harvard University Press, 1993). See Latour, *Irréductions,* 263. [Please note that the excerpt appearing here differs from the published translation.—*Trans.*]

15. This celebration entered the classical narrative via Gödel's theorem, the mathematical death of the hope of a formal language capable of determining the truth value of any statement in that language. But I'll resist the temptation to attribute to Gödel everything that populates this narrative. Dead ends, impossibilities, and paradoxes each arise from the practices that produce them. It is, first and foremost, the narrative that glorifies them that creates a "common ground" for them, a place—and the multiple references to Gödel bear this out—we tend to enter as if it were a mill.

7. THE POWER OF PHYSICAL LAWS

1. Steven Weinberg, *Dreams of a Final Theory* (London: Vintage Books, 1993), 36.

2. Patricia S. Churchland, "Is Neuroscience Relevant to Philosophy," *Canadian Journal of Philosophy*, supplementary volume 16 (1990): 341.

3. Bruno Latour, *Petite Réflexion sur le culte moderne des dieux faitiches* (Le Plessis-Robinson: Synthélabo, "Les Empêcheurs de penser en rond," 1996).

4. Isabelle Stengers, *The Invention of Modern Science*, trans. Daniel W. Smith (Minneapolis: University of Minnesota Press, 2000).

5. Note that this expectation can cooperate in the creation of a rather powerful subjectivity. The terrible excesses of what is known as "animal experimentation," rightly denounced as "torture," reflect, or so it would seem, in terms of the self-blinding of the "experimenter" with regard to her work, the effects of this dynamic of blind accumulation, where all observations are equally valid. For the sake of a future when we will finally understand, every observation seems worthy of being carried out—who knows, it may even turn out to be important one day.

8. THE SINGULARITY OF FALLING BODIES

1. Gottfried Wilhelm Leibniz, *Die Philosophischen Schriften*, vol. 3, ed. C. I. Gerhardt (Hildesheim: Georg Olms, 1971), 43–44.

2. See Wilson L. Scott, *The Conflict between Atomism and Conservation Theory, 1644–1860* (London: Macdonald, 1970).

3. Some of Newton's followers, like Reverend Clarke, rejected Leibniz's claims by turning to Catelan's demonstration. See *Correspondance Leibniz-Catelan*, with an introduction by A. Robinet (Paris: Presses Universitaires de France, 1957), 200–203. The argument appears in a note written in 1717 after Leibniz's death. Fortunately for Clarke, Leibniz was no longer around to reply.

4. See Isabelle Stengers, "Les affaires Galilée," in *Éléments d'histoire des sciences*, ed. Michel Serres (Paris: Bordas, 1989), 223–49.

5. Concerning Leibniz's physics, see Martial Guéroult, *Leibniz, Dynamique et métaphysique* (Paris: Aubier-Montaigne, 1967).

6. Stengers, "Les affaires Galilée," 98–99.

9. THE LAGRANGIAN EVENT

1. Clifford Truesdell, *Six Lectures on Modern Natural Philosophy* (Berlin: Springer Verlag, 1966), 85.

2. Clifford Truesdell, "Statistical Mechanics and Continuum Mechanics," in *An Idiot's Fugitive Essays on Science: Methods, Criticism, Training, Circumstances* (New York: Springer Verlag, 1984), 75.

3. Which Truesdell introduced when he became a historian of "pre-Lagrangian" mechanics, that is to say, a pitiless critic of all those who had written that history by taking their inspiration directly from the historical presentation provided by Lagrange, which, obviously, led directly to him.

4. See Book I, "The Science Wars."

5. It is important not to confuse "live force," which is a state function, with the force used here to evaluate instantaneous effect, a force that determines acceleration. Live force is, in itself, a measurement of what a body has "gained" or is likely to "lose" as a result of its motion. The force acting from moment to moment, however, presents us with the problem of measurement. Naturally, the two work together. Galileo, in his *Discourses and Mathematical Demonstrations Relating to Two New Sciences*, had to make use of the motion of a pendulum, which exhibited the conservation of cause and effect, to illustrate the principle that "the degrees of velocity that a given moving body acquires on different inclined planes are equal, providing that the heights of the planes are equal." But in 1639 he dictated to his student Viviani a new demonstration, which he thought would "further confirm the truth of the principle we have already tested earlier using reasonable arguments and experiment." And it was this demonstration that introduced the measurement of "degree of velocity" gained *at each moment* through the equilibrium between two bodies. For Leibniz, the principled nature of the measurement of instantaneous acceleration through equilibrium is reflected in the fact that it is precisely the point at which the two definitions of live force—physical, which is to say measurable, and metaphysical—are distinguished. From the physical point of view, equilibrium measures *dead* force, and live force is defined as the integral over space of the dead forces: it appears to be determined by its interaction with other bodies that may, or may not, serve as an obstacle to its growth through the accumulation of dead forces. But from the metaphysical point of view, the variation of live force reflects the spontaneous, autonomous development of an activity possessing its own internal law, the law of monads. The reversibility of these points of view—the variation of live force seen as being entirely determined by the universal "conspiracy" of external causes, or as entirely spontaneous, *causa sui*, and therefore not susceptible to measurement (to measure something, you have to interact with it)—constitutes the direction of Leibniz's thought.

6. See Betty J. Dobbs, *The Janus Face of Genius* (Cambridge: Cambridge University Press, 1991).

7. Part of the Lagrangian redefinition is supposed to be valid for all forces, not only conservative forces, which are subject to d'Alembert's principle. Of course, in this case, the formalism loses most of its power.

8. Bertrand Russell, "On the Notion of a Cause," in *Mysticism and Logic* (New York: Longmans, Green and Co., 1919), 180.

9. In the classical models of mathematical economics, this determination is expressed by a handful of readily endorsed restrictions, which amount to subjecting economics to d'Alembert's principle . . . and in so doing stripping it of any hope of relevance.

10. The relationship between the redefinition of equilibrium and work

is the major contribution made by statics to dynamics. For this aspect of the question, see Pierre Duhem, *L'évolution de la mécanique* (Paris: Vrin, 1992).

11. The very strange figure of economic life portrayed by the "market" in mathematical economics, with its fully informed and amnesiac actors, each acting for itself in a context of strict equality with respect to momentary supply and demand, is nothing other than an extremely violent "rationalization" that transforms economy into a science of state functions. Economic "rationality" is not a conquest, and is in no way justified by a distinction between what is "rational" and what is not. The equality among "actors" is required; it is an inherent condition of what has become a veritable machine for creating equivalences made possible by the use of Lagrangian equations.

10. ABSTRACT MEASUREMENT

1. Quoted by François Vatin, *Le Travail: Économie et physique, 1780–1830* (Paris: Presses Universitaires de France, 1993), 74.

2. See Isabelle Stengers and Didier Gille, "Time and Representation," in *Power and Invention* (Minneapolis: University of Minnesota Press, 1997), 177–211.

3. Here, we should note the exception of the Soviet school of mathematics, which, in the twentieth century, was a pioneer in the study of nonlinear dynamics. See Simon Diner, "Les voies du chaos déterministe dans l'école russe," in *Chaos et déterminisme* (Paris: Le Seuil, 1992), 331–70.

4. In More *Heat Than Light: Economics as Social Physics, Physics as Nature's Economics* (Cambridge: Cambridge University Press, 1989), 242–48, Philip Mirowski describes the initial misunderstanding that took place between 1898 and 1900, pitting Harmann Laurent, a mathematician specializing in rational mechanics, against Walras, and then Pareto. Laurent, who was very interested in the problem, asked the economists how they identified their "unit of value." This, in fact, corresponded to the question of finding out if "economic forces" were conservative, for the possibility of determining a value reflected the existence of a state function. Walras and Pareto were never willing to consider this "minor" problem as anything more than mathematical hairsplitting. In *Complexity, the Emerging Science at the Edge of Order and Chaos* (New York: Simon & Schuster, 1992), M. Mitchell Waldrop describes the tumultuous meeting between contemporary physicists and economists at the Santa Fe Institute. The following is a typical example of an exchange about economic assumptions: "You guys really *believe* that?" one of the physicists asked. The economists replied, "Yeah, but this allows us to solve these problems. If you don't make these assumptions, then you can't do *anything.*"

11. HEAT AT WORK

1. Donald Cardwell, in *From Watt to Clausius* (London: Heinemann, 1971), 180–81, notes that the improvement in output of the English engines owed very little to an empirical discovery of the importance of criteria similar to those Carnot would define as "optimal output" and a great deal, as Poisson had suspected, to the perfection of their technical construction.

2. Sadi Carnot, *Reflections on the Motive Power of Heat and on Machines Fitted to Develop This Power,* trans. R. H. Thurston (Baltimore: Waverly Press, 1943), 56 [emphasis in the original.—*Trans.*].

3. Note that this is where, providentially, Carnot's "twofold error" is found (see Cardwell, *From Watt to Clausius*). He must assume that, to be able to return to its initial state, all the caloric received during isothermal expansion has to pass to the cold source during isothermal compression (which *we consider* to be false). But he also believes that the same quantity of caloric that "falls" one degree in temperature produces even more work when the operation occurs at a lower temperature, which is also false for us but necessarily follows from another "experimental fact" accepted by all Carnot's contemporaries: the increase of the specific heat of a body with its volume. This "fact" will disappear along with the identification of heat with a fluid that is conserved. It is because the effects of these two "errors" reciprocally compensate each other in the formula for optimal output established by Carnot that the formula has, in a most improbable manner, resisted the abandonment of the caloric theory of heat.

4. Carnot, *Reflections on the Motive Power of Heat and on Machines Fitted to Develop This Power,* 58.

12. THE STARS, LIKE BLESSED GODS

1. More specifically, the result of work by Hamilton and the German mathematician K. J. G. Jacobi, who, in 1866, gave the "Hamilton-Jacobi" equations their final form.

2. In Lewis Campbell and William Garnett, *The Life of James Clerk Maxwell* (London: Macmillan, 1882), 647–48.

3. In "phase space," a $6N$-dimensional space, each corresponding to one of the $6N$ independent variables of a dynamic system, every state of this system can be represented by a point, and each trajectory by a curve. Whenever the variables are cyclical, it is as if the various components of the system, each characterized by a position coordinate–moment coordinate pair, mv (then called an "action variable" and symbolized as J), do not interact. This means that the phase space can be broken down into $3N$ spaces of two dimensions each. Within each of those subspaces, the trajectory of the system describes a curve that is closed in on itself, whose velocity (constant since there is no longer any interaction) corresponds to a frequency and whose position coordinate corresponds to the angle of rotation. This rep-

resentation may seem counterintuitive, but astronomers find it compellingly beautiful for it immediately yields Kepler's law, according to which the radius that unites the Sun to a planet sweeps across constant surfaces over constant time periods. This is another way of saying that J is invariant over time.

4. That is why this "technical" exploration is crucial for the remainder of my argument: in this case, the question of integrability is central to the "renewal of dynamics" proposed by Ilya Prigogine.

5. For this and the following, see J. Laskar, "La stabilité du système solaire," in *Chaos et déterminisme* (Paris: Le Seuil, 1992), 170–211.

6. The Hamiltonian of the system being studied is written $H = H0 + \lambda V$. Here, λ corresponds to the "coupling coefficient," with a value less than 1 (the smaller its value, the greater the system being studied is "similar to" the reference system). The introduction of the coupling coefficient indicates that what is being studied is not a system but a class of perturbed systems, a class that resembles all the systems characterized by a perturbation of the same form (defined by V) but of variable intensity (quantified by λ). Correlatively, the sought-for action variables, J', are defined by a series development of the form $J + \lambda J1 + \lambda 2 J2 + \ldots$ which ensures a monotonic decrease of the terms, because the coupling coefficient λ is less than 1. So, whenever this coefficient tends toward zero, J' tends toward J, the action variable of the nonperturbed system.

7. See J. Laskar, "Le problème des petits dénominateurs," in *Chaos et déterminisme*, 188–93.

14. THE THREEFOLD POWER OF THE QUEEN OF HEAVEN

1. Edward E. Daub, "Maxwell's Demon," *Studies in the History and Philosophy of Science* 1 (1970): 213–26. The quotation appears on page 220.

2. From the article titled "Diffusion" in the *Encyclopaedia Britannica*, 9th ed.; quoted in ibid., 217.

3. That is, whenever this procedure culminates in the welcome result of transforming the syntax of the description by enabling a transition from "laws" that confirm conservation to "dissipative" experimental magnitudes. See Nancy Cartwright, *How the Laws of Physics Lie* (Oxford: Oxford University Press, 1983). I'll return to Cartwright's book in "Quantum Mechanics: The End of the Dream."

4. See Book II, "The Invention of Mechanics: Power and Reason."

5. See Book I, "The Science Wars."

6. The "well-behaved" nature of the power attributed, through Laplace's demon, to the laws of motion is found in the ease with which relations of resemblance later shown to be indefensible were accepted at the time. Thus, the concept of equilibrium was for a long time considered to be "common" to dynamics and the other sciences, which also described the evolution

"toward equilibrium" of whatever concerned them: chemical, electrical, or thermal phenomena primarily. But this overlooked what could no longer be overlooked when Maxwell's Queen of Heaven affirmed her power: a "pure" dynamic system never evolves "toward" equilibrium. For example, only friction, which interferes with movement, is responsible for causing a pendulum to finally come to a standstill in a state of motionless equilibrium.

15. ANAMNESIS

1. Ernst Mach, "Die Leitgedanken meiner naturwissenschaftlichen Erkenntnislehre und ihre Aufnahme durch die Zeitgensossen," *Physikalische Zeitschrift,* vol. 11 (1910): 599–606. See Ernst Mach, *Physical Reality: Philosophical Essays on 20th Century Physics,* ed. Stephen Toulmin (New York: Harper and Row, 1970).

2. See Book V, "The Arrow of Time: Prigogine's Challenge."

3. Gilles Deleuze and Félix Guattari, *What Is Philosophy?,* trans. Hugh Tomlinson and Graham Burchell (New York: Columbia University Press, 1994), 124.

4. Ibid., 125.

16. ENERGY IS CONSERVED!

1. Thomas Kuhn, "Energy Conservation as an Example of Simultaneous Discovery," in *The Essential Tension: Selected Studies in Scientific Tradition and Change* (Chicago: University of Chicago Press, 1977).

2. See Yehuda Elkana, *The Discovery of the Conservation of Energy* (London: Hutchinson, 1974).

3. See Loup Verlet, *La Malle de Newton* (Paris: Gallimard, 1993).

4. Friedrich Engels, *The Dialectics of Nature,* trans. J. B. S. Haldane (New York: International Publishers, 1960), 28.

5. In this way Engels, using other means, resumed Hegel's battle against mechanics. Mechanics, because it is the science of an object that determines its own categories, defining cause and effect in its own terms, was for Hegel a threat to philosophy. It seemed to be capable of thinking the truth of motion for itself and by itself. In section 3 ("Measure") of *The Science of Logic,* Hegel contrasted the difference between mechanical measurement, which merely specifies, and chemical measurement by affinity, which is associated with "real" measurement. In the conservation of energy, Engels found an opportunity for a different strategy: the very triumph of mechanical measurement, of work as measurement, precisely because it equates qualitatively different forms of "motion," signals its inability to theorize what it measures. This is the theory that was supposed to be presented in the second, unfinished, part of *The Dialectics of Nature.* See Éric Alliez and Isabelle Stengers, "Énergie et valeur: le problème de la conservation chez Engels et Marx," in *Contre-temps: Les pouvoirs de l'argent* (Paris: Éditions Michel de Maule, 1988).

6. Having done so, Helmholtz, in this new context in which heat had become a quantity, returned to the interpretation that Leibniz, in his fifth letter to Clarke, gave to the loss of movement that occurs when two "soft" or nonelastic objects strike one another. Are active forces lost? "The author objects, that two soft or un-elastic bodies meeting together, lose some of their force. I answer, no. 'Tis true, their wholes lose it with respect to their total motion; but their parts receive it, being shaken [internally] by the force of the concourse. And therefore that loss of force, is only in appearance. The forces are not destroyed, but scattered among the small parts. The bodies do not lose their forces; but the case here is the same, as when men change great money into small" (*The Leibniz-Clarke Correspondence: Together with Extracts from Newton's Principia and Opticks* [Manchester: Manchester University Press, 1998], 87). We should not forget, however, that, for Leibniz, the philosophical truth of "active" or "live" force is not mechanical. In fact, with regard to Helmholtz, the question is also complex, Kant's influence being very much present, for he accepts the "Kantian" obligation, universalizing the categories made explicit by mechanics to all phenomena. Only the English scientists, like Joule and Thomson, can be said to be "truly" realists. When they speak of force or energy, it is the world created by god and not the phenomena we know rationally that is in question.

7. This difference is one of the central themes of "The Invention of Mechanics: Power and Reason."

8. See "The Invention of Mechanics." I won't repeat the description of the cycle here but will limit myself to reminding readers that it functioned by using heat taken from a hot source and given to a cold source to produce mechanical work. The principle of its operation satisfies one need and one alone: in order to optimize the yield of the cycle, all the transformations had to be reversible; that is, they had to avoid, through some form of perfect control, the direct flow of heat between two bodies at different temperatures.

17. THE NOT SO PROFOUND MYSTERY OF ENTROPY

1. All of Clausius's articles on thermodynamics have been collected in *Théorie mécanique de la chaleur,* recently reissued by Jacques Gabay (Sceaux, 1991). See *The Mechanical Theory of Heat,* trans. Walter R. Browne (London: Macmillan, 1879).

2. In doing so, Clausius destroyed with a single blow what for Thomson had been the greatest advantage of the Carnot cycle: the central role played by the "latent heat of expansion," the heat absorbed whenever a gas expands. The existence of this latent heat of expansion was a direct consequence of the caloric theory: whenever a gas moves a piston without any exchange of heat (adiabatic expansion), its temperature decreases. However, according to the caloric theory, temperature is a function of the part of caloric that is not

"absorbed" by the body, the way water is absorbed by a sponge, its "free" part rather than its "latent" part. Therefore, a drop in temperature that is not associated with an exchange of heat with the environment measures the transition, in latent form, of part of the caloric present in the body, that is, an increase in the specific heat of the body, its "capacity" to absorb heat. And because Carnot believed that the specific heat of a gas increased with its volume, he was able to conclude that the drop in caloric should produce even more motive power than it did at lower temperatures. Moreover, the latent heat of expansion is, according to Clapeyron's reading of the Carnot cycle, the key magnitude in measuring the output of thermal machines. This immediately drew the attention of William Thomson for, in the 1840s, measurement of this quantity had become the Holy Grail of experimental physics (see D. S. L. Cardwell, *From Watt to Clausius: The Rise of Thermodynamics in the Early Industrial Age* [London: Heinemann, 1971]). Yet, when Clausius reinterpreted the cycle in the context of the conservation of energy, this key magnitude was the first victim of the "new way of seeing" he proposed. This latent heat, supposedly "hidden from our perception," "*does not exist at all,*" claimed Clausius, "it is *consumed* as work during changes of state."

3. The little-known story of how Thomson and Tait, who affirmed Thomson's priority over Clausius, were flagrantly caught propounding an absurdity in 1879, precisely because they believed they could provide a general definition of this loss, is described in Edward E. Daub, "Entropy and Dissipation," *Historical Studies in the Physical Sciences* 2 (1970): 321–54. See also Cardwell, *From Watt to Clausius.*

4. To make things a bit clearer, we can also say that Clausius helped make the Lagrangian fiction intelligible by defining the first physical–mathematical state function "for itself" (energy was already the central quantity in mechanics before Lagrange). In this sense, his work is of great educational value. Similarly, he must have recognized, and had to explain, the difference between a path as understood by Lagrange and by Carnot. All the variables in a mechanical system are free, whereas, in the case of "thermodynamic pathways," the variation must be limited to a couple of variables, the others (quantity of heat for adiabatic systems, temperature for isothermal systems) being kept constant. Pierre Duhem, who continued Clausius's work, is also the one who presented, far better than anyone else, the "Lagrangian fiction" that other authors have neglected as being simply a means in the service of the mechanical description of systems. Taking my inspiration from Duhem in discussing the "Lagrangian event" in "The Invention of Mechanics," I described mechanics from a "thermodynamic" point of view.

5. Clausius postulated that the cycle, in addition to the heat it absorbs from the hot source, absorbs an amount of heat equal to the heat it will give up to the cold source from a third source at some intermediate temperature.

The transactions take place "as if" all the heat absorbed at the hot source was converted into work, while all the heat absorbed at the intermediate source flowed to the cold source. Obviously, an operation such as this only has meaning in terms of compensation: the idea that it is the heat absorbed at the hot source that is converted into work does not correspond to any physical hypothesis.

6. The adiabatic curves make no explicit contribution, but they are perfectly defined because the change of temperature corresponding to the transition between one isothermal curve and another is sufficient to identify the adiabatic curve that had to have been followed to realize that transition.

7. The definition of Q/T as a state function duplicates the original simplicity of the cycle. From the point of view of the theory of heat-caloric, if Q and T are taken as our variables, the cycle can be represented as a rectangle; it loses as much heat at the cold source as it gains at the hot source. And we can also represent the ideal Clausius cycle as a rectangle if our variables are Q/T and T. But the similarity ends there. The conservation of Q was a consequence of the nature of heat, while the key role of the function Q/T is an integral part of the concept of reversible transformation invented by Carnot and Clausius.

8. The "rational" measurement of a dissipative process then implies the elimination of the dissipation that characterizes it. This is what Carnot accomplished for heat and work transformations and what chemical thermodynamics would bring about for processes involving a chemical reaction: all the thermodynamic magnitudes characterizing a chemical reaction are in fact directed at a "transition" from one equilibrium state to another that rationally mimics the spontaneous reaction.

9. The fact that this production is positive by definition has to do with the definition of entropy. If another definition had been used, it might have been negative. The important thing is that it should have a known sign.

10. For Berthelot, just as mechanical equilibrium was defined by minimum potential energy, chemical equilibrium could be defined as the point when the "chemical" energy in the reaction milieu was at a minimum. Chemical equilibrium could be obtained when the maximum "chemical work" was achieved, this work being measured by the heat of reaction given off by the system.

11. The entropy maximum can be used to define the equilibrium states achieved by an isolated thermal system because, if no exchange of heat has taken place, any change in dQ will be associated with uncompensated irreversible processes. But in other thermodynamic systems, changes are defined by other limit conditions. For example, an irreversible transition to equilibrium can occur at constant temperature, which implies heat exchange with the milieu at constant pressure (imagine a chemical reaction taking place in a water bath at the boiling point of water and under atmo-

spheric pressure). In this case, it is not the entropy, *S*, but the thermodynamic potential, *G*, often called "Gibbs free energy," that reaches its extreme value (minimal rather than maximal) at equilibrium.

18. THE OBLIGATIONS OF THE PHYSICIST

1. Although some creators, such as William Thomson, were helping to transform the world and participate in the construction of empire. See Crosbie Smith and M. Norton Wise, *Energy and Empire: A Biographical Study of Lord Kelvin* (Cambridge: Cambridge University Press, 1989).

2. Henri Poincaré, *La Science et l'hypothèse* (Paris: Flammarion, 1906), 158.

3. Pierre Duhem, *The Aim and Structure of Physical Theory*, trans. Philip P. Wiener (Princeton, N.J.: Princeton University Press, 1954), 390.

4. This is the primary difference between Pierre Duhem and Karl Popper, who will use the idea of logic's inability to ensure that a theory can be refuted against the positivists, in order to prevent a theoretical modification that might invalidate "experimental contradiction." Popper went on to transform this topic into a directive, obligating those who claim to be scientists to take an initiative that logic does not require of them. In doing so, he would overlook one "small problem": the contradiction, whether it is silenced by means of a "conventionalist stratagem" or whether it is given the power of refutation, only makes sense in the theoretical-experimental field.

5. Albert Einstein, "The Principles of Scientific Research," in *The World as I See It* (New York: Philosophical Library, 1949).

6. Abel Rey, *La théorie de la physique chez les physiciens contemporains* (Paris: Alcan, 1923), 211–12. Rey was also the author of "La philosophie scientifique de M. Duhem," which appeared in the *Revue de métaphysique et de morale*, 12th year (July 1904): 699, in which he connects Duhem's "scientific skepticism" and his Catholic faith. Duhem's response, "The Physics of a Believer," and his commentary on *La théorie de la physique chez les physiciens contemporains*, "The Value of Physical Theory," appear in the appendix to the revised edition of *The Aim and Structure of Physical Theory*, trans. Philip P. Wiener (Princeton, N.J.: Princeton University Press, 1991).

7. J. von Liebig, *Lord Bacon* (Paris: Baillère et Fils, 1894).

8. See W. Paul, *The Sorcerer's Apprentice: The French Scientist's Image of German Science, 1840–1919* (Gainesville: University of Florida Press, 1972), and Pierre Duhem's astonishing *German Science: Some Reflections on German Science: German Science and German Virtues* (La Salle, Ill.: Open Court, 1991).

19. PERCOLATION

1. Michel Serres, *The Origins of Geometry* (Manchester: Clinamen Press, 2001), 43.

2. See Book IV, "Quantum Mechanics: The End of the Dream."

3. In 1871, the quarrel about priority to which Maxwell alluded in the quote at the beginning of chapter 14 pitted the young Boltzmann, who, in 1866, had published a "proof" that the second law is nothing but a mechanical theorem, against Clausius, who had recently published a similar demonstration.

4. William Thomson, "Kinetic Theory of the Dissipation of Energy," *Nature* 9 (1874): 442; quoted in Edward E. Daub, "Maxwell's Demon," *Studies in the History and Philosophy of Science* 1, no. 3 (1970): 216.

5. See Pierre Duhem, *Mixture and Chemical Combination and Related Essays,* ed. and trans. Paul Needham, Boston Studies in the Philosophy of Science, 223 (Dordrecht: Kluwer Academic Publishers, 2002).

6. The possibility of characterizing a population of unstable particles with a lifetime, that is, by an exponential law of decay (or de-excitation for excited atoms), signifies that every member of this population possesses, a priori, and regardless of the physical environment, the same probability of decay at every instant. An event characterized by a lifetime cannot, by definition, be influenced by variables such as those Maxwell's demon is able to manipulate.

7. For the time being, I'm leaving aside the question of the quantum interpretation of the event. The irony of history is that, by the time the quantum event came to pose a problem, the generation of scientists interested in thermodynamic irreversibility had disappeared, and the increase in entropy was seen as a problem that had been solved. In "Thermodynamique et cinétique: les deux sources non classiques de la théorie de la relativité restreinte" (in *Nouvelles Tendances en histoire et philosophie des sciences* [Brussels: Académie des sciences des lettres et des beaux-arts de Belgique, 1993]), Yves Pierseaux demonstrates young Einstein's uniqueness in this regard. In 1905, Einstein's published work concerned Brownian motion, Max Planck's quanta, and relativity; Einstein never considered the second law of thermodynamics to be a "simple matter of probability." For Einstein, probabilities should allow us to understand the second law, not resolve the problem. It's even possible, as Pierseaux notes, that the discrete quanta of Max Planck ushered in the possibility for Einstein of a "kinetics" of motion for which special relativity might supply certain premises. Following this assumption, Einstein himself would have been redefined by the meaning embedded in his work when it became a synonym for a "revolution" affecting the science (dynamics) of space, time, and motion, defining this as "the" science, one that had remained unified, invariant, and omnipotent ever since Newton and which required a revolution to challenge.

8. See Thomas S. Kuhn, *Black-Body Theory and the Quantum Discontinuity: 1894–1912* (Chicago: University of Chicago Press, 1987), 60–71.

9. From this follows the recurrent, but bizarre, notion that equilibrium is

not at all characterized by uniformity or the overlooking of differences. On the contrary, it is the state to which the greatest microscopic diversity corresponds. The idea is bizarre because of the hypothesis of equal probability, which renders each of those various states insignificant. Regardless of the extraordinary microscopic configuration occurring at a given moment, its extraordinary character does not "count" either for us, ordinary macroscopic observers, or for anything that might invent consequences for such a configuration (for example, with respect to the origin of life). In effect, this configuration is by definition transitory and not reproducible.

10. See Book V, "The Arrow of Time: Prigogine's Challenge."

11. The only macroscopic behavior probability truly favors is the state of stable equilibrium. If a system has been prepared in a state of equilibrium, the overwhelming majority of dynamic changes, whether they are caused by a reversal of velocities or not, will maintain it at equilibrium. In other words, probability allows us to characterize the equilibrium state but not to "save" irreversible change. Its field is, therefore, the same as that of thermodynamic potentials, whose end point alone, corresponding to an equilibrium state, is defined.

20. IN PLACE OF AN EPILOGUE

1. "La physique des électrons," in Paul Langevin, *La Physique depuis vingt ans* (Paris: Doin, 1923).

2. See Yves Pierseaux, "Thermodynamique et cinétique: les deux sources non classiques de la théorie de la relativité restreinte," in Langevin, *La Physique depuis vingt ans*.

3. For this and the following comments, see Thomas S. Kuhn, *Black-Body Theory and the Quantum Discontinuity: 1894–1912* (Chicago: Chicago University Press, 1987).

4. The "subject" in the sense of what is now important to physicists. It is important to keep in mind that Max Planck himself maintained an original position until the end. For Planck, the probabilistic interpretation pointed to other things because, for it to play a role, it was necessary to add to the dynamic model of Boltzmannian collisions and the electrodynamics of radiation an additional hypothesis, foreign to dynamics: the assumption of "elementary disorder" (molecular chaos for Boltzmann, "natural" radiation for Planck). In his memoirs, Planck emphasized that it was this assumption that ensured the relevance of the intelligible laws of dynamics in cases where, as he had maintained previously, nature is not "indifferent," when natural changes reflect "preferences," which their irreversibility acknowledges. In other words, the unification brought about by dynamics is creative. We could say that it creates the intelligible unity of what it "captures" in the net of its reversible equivalences, but it also creates the additional hypotheses each model necessitates. Those hypotheses are important, therefore. They are what point to, in spite of the assumed indifference of the model, the preferences that irreversibility signifies.

INDEX

Trained as a chemist and philosopher, **Isabelle Stengers** has authored or coauthored more than twenty-five books and two hundred articles on the philosophy of science. In the 1970s and 1980s, she worked with Nobel Prize recipient Ilya Prigogine, with whom she wrote *Order out of Chaos: Man's New Dialogue with Nature.* Her interests include chaos theory, the history of science, the popularization of the sciences, and the contested status of hypnosis as a legitimate form of psychotherapy. She is a professor of philosophy at the Université Libre de Bruxelles. Her books *Power and Invention: Situating Science* (1997) and *The Invention of Modern Science* (2000) have been translated into English and published by the University of Minnesota Press.

Robert Bononno has translated more than a dozen books, including *Psychoanalysis and the Challenge of Islam* by Fethi Benslama (Minnesota, 2009) and *Decolonization and the Decolonized* by Albert Memmi (Minnesota, 2006). He was a finalist for the French–American translation prize for his translation of René Crevel's *My Body and I.*